AF251673

# The life of
the lyrebird

# The life of
# the lyrebird

L. H. Smith

**William Heinemann Australia**

First published 1988 by
William Heinemann Australia
(a division of Heinemann Publishers Australia Pty Ltd)
85 Abinger Street, Richmond, Victoria 3121



Designed by Susie Stubbs
Maps by Chris Haddon
Marbled endpapers by Margo Snape
Typeset in Garamond Book ITC
by MacKenzie Typesetting
Printed by Mandarin Offset, Hong Kong

National Library of Australia
cataloguing-in-publication data:

Smith, L.H.
    The life of the lyrebird.
    ISBN 0 85561 122 7.
    1. Lyrebirds. I. Title.
598.8

The publisher wishes to thank the Edward A. Green
Charitable Foundation for its generosity in subsidizing
the production of this book.

# Contents


Illustrations ix

Preface xiii

Acknowledgements xv

1 Historical introduction 1

2 General observations on the superb lyrebird 13

3 The male superb lyrebird 26

4 The female superb lyrebird 39

5 The lyrebird's breeding cycle 57

6 The young lyrebird 63

7 The lyrebird's tail 79

8 The song of the lyrebird 94

9 Lyrebirds in captivity 103

10 Lyrebirds in Tasmania 111

11 Lyrebirds in Queensland 125

References 139

Index 143

# List of illustrations


## Black and white photographs

| | Page |
|---|---|
| The first painting of a superb lyrebird | 4 |
| An engraving by Giacomelli | 5 |
| Spotty finds a worm | 16 |
| Timothy, displaying in the firebreak | 18 |
| Spotty, stretching his wings | 21 |
| Spotty reacted instantly to a message | 23 |
| The head of the superb lyrebird | 27 |
| The legs and claws of a mature male superb lyrebird | 28 |
| Spotty courting a female lyrebird | 30 |
| Spotty in a 'wing-raising' display | 31 |
| Spotty pauses in his quest for food | 34 |
| A yellow robin was accidentally caught in Spotty's claw | 35 |
| An adult female lyrebird | 40 |
| A female lyrebird giving a threat display | 42 |
| As a prelude to mating the female goes under the tail of the male | 43 |
| The nest of the superb lyrebird | 44 |
| A female lyrebird with a bent tail | 46 |
| A female lyrebird carrying nesting material | 48 |
| A female lyrebird flying away from her nest | 50 |
| A female lyrebird pauses before springing across to feed her chick | 52 |
| The lyrebird uses his two large claws to remove vegetation | 61 |
| The underside of the tail of a lyrebird chick | 65 |
| A female lyrebird depositing the chick's dropping in a stream | 67 |
| A female lyrebird enticing her chick from the nest | 69 |
| The author's seven-year-old daughter watching a lyrebird chick | 74 |
| A confrontation between Red/Blue and another immature lyrebird | 76 |
| The mutual interaction of young lyrebirds | 77 |
| The detailed structure of the distal ends of a mature and of an immature male lyrate respectively | 81 |
| An immature lyrebird displaying | 85 |
| Red/Blue grew his first filamentary feather in his fourth year | 86 |

The first four generations of lyrates     87
Later generations of lyrates     88
The medians change in size and structure as the bird grows older     89
Single Red at an early stage of his sixth tail moult     90
Single Red at the conclusion of his sixth tail moult     91
White/White growing a new tail early in his sixth year     92
A mature male imitating the black cockatoo     99
Two female lyrebirds in the Adelaide Zoo in mutual display     105
Our 'tame' lyrebird, drinking from a glass     107
Our lyrebird being weighed     109
The Albert lyrebird displaying     129
A fern gully in Sherbrooke Forest     134


# Colour plates


Between pages

Collins's colour print of a painting by a 'competent artist'     6-7
G.J. Broinowski's plate of the superb lyrebird     6-7
A mature male lyrebird singing on a large log     6-7
The gray (or crested) gallito     6-7
Gould's painting of the superb lyrebird     6-7
The dense vegetation of Sherbrooke Gully     22-23
A mature male superb lyrebird in tail moult in mutual display with an immature lyrebird     22-23
Spotty in his favourite bathing pool     22-23
A lyrebird squeezing water out of his tail feathers after bathing     22-23
Spotty removing water from feathers under his chin     22-23
A female lyrebird preening after her bath     22-23
The lyrebird using the weight of his body to flatten bracken     22-23
A side view of the 'invitation' display     38-39
A frontal view of the 'invitation' display     38-39
A mature male lyrebird attempting to mount an immature male     38-39
A mature male superb lyrebird courting a female     38-39
A mature male lyrebird engaged in pre-nuptial courtship     38-39
Spotty giving his territorial call     38-39
A mature male lyrebird displaying, facing the camera     38-39
A rear view of the display     38-39
A lyrebird taking a drink     54-55
A mature male lyrebird displaying, his body concealed     54-55
An immature male lyrebird, displaying to a fern     54-55
The conclusion of a display     54-55
A mature male lyrebird and an immature male, engaged in a fight     54-55
A mature male lyrebird feeding his four-month-old chick     70-71
Cradle of a lyrebird's nest     70-71
A lyrebird's nest in Sherbrooke Forest     70-71
Lyrebird eggs from different districts     70-71
A female lyrebird settling down on her egg     70-71
A lyrebird chick aged nine days     70-71
A lyrebird chick aged forty-four days     70-71
A lyrebird chick aged thirty-nine days, temporarily outside the nest     86-87
A lyrebird chick aged about two and a half weeks, showing the wing feathers emerging     86-87
A female lyrebird with food for her chick     86-87
A female lyrebird feeding her chick     86-87
A female lyrebird, with her chick's dropping in her beak     86-87
This chick stood on the platform of the nest for two hours     86-87
A female lyrebird and her six-month-old chick feeding side by side     86-87

A female lyrebird feeding her
    six-month-old chick     86-87

An immature male lyrebird aged ten
    months     102-103

A young male lyrebird in his fourth
    year     102-103

A mature male lyrebird growing a
    new tail     102-103

This bird had acquired eleven
    filamentaries by his eighth year     102-103

The common fault bars in feathers     102-103

Two immature male lyrebirds
    displaying on the same mound     102-103

Spotty, on a moss-covered log, pauses
    in his song     102-103

A view of Pandanni Flat, Mount Field
    National Park, Tasmania     118-119

A lyrebird mound in a bed of
    spaghum moss in Tasmania     118-119

The kookaburra     118-119

The crimson rosella     118-119

The golden whistler (male)     118-119

The grey thrush     118-119

The red wattlebird     118-119

The yellow robin     118-119

The satin bowerbird at his bower     118-119

The grey-fantail     118-119

Gould's original plate of the Albert
    lyrebird     118-119


# Maps


Map 1    Lyrebirdmen's journeys north-
    west of Parramatta, 1798     3

Map 2    Map of Sherbrooke Forest     12

Map 3    Lyrebirds in Tasmania     122

Map 4    Distribution of Albert and
    superb lyrebirds in north-east
    New South Wales and
    south-east Queensland     126

Map 5    Distribution of lyrebirds in
    north-east New South Wales
    and south-east Queensland     128

# Preface

This is my story — a story of the discovery of the lyrebird 180 years ago, of the interest its discovery aroused in the scientific world, and of its shameful treatment by sportsmen and collectors, almost to the point of extinction. It is a story of the efforts of a small band of nature-lovers who strove to arrest the forces of destruction and whose success was marked by the elevation of the lyrebird to the position of a world celebrity, its great attractions being a song combining appealing musical quality with unsurpassed mimicry, and a spectacular display of its elaborate tail feathers. It is a story of painstaking studies on the behaviour of the lyrebird — courtship rituals, family relationships and, in more recent times, of attempts to unravel the mysteries of the remarkable song of the lyrebird.

In much of this work, Sherbrooke Forest has played a major role. It was from Sherbrooke that the lyrebird's song first echoed around the world, just over fifty years ago. But now the forest no longer resounds to the song of the Master of Mimicry. Why?

The very silence of Sherbrooke, hanging over the forest like a pall, challenges us to seek an answer. The answer is, I think, that, however unwittingly, Sherbrooke's famous songster has been sacrificed on the altar of tourism and real-estate. It is not so much the ever-increasing numbers of day-picnickers who use the tourist facilities provided by the government over the years, or the increasing numbers of houses around the perimeter of the forest, that have in themselves destroyed the lyrebird, but rather that these developments have precluded the management of the forest to ensure that it continued to provide a home and a livelihood for the lyrebird. The gradual accumulation of forest litter, coupled with the proliferation of noxious weeds and native plants such as ferns and wire-grass, has, in the end, reduced the capacity of the forest to support a substantial lyrebird population; while those birds which do endeavour to make the forest their home are under constant threat from predators such as dogs, cats and foxes.

It has come to be recognized that other Australian species need special ecological environments. Leadbeater's possum thrives in re-growth forest; the needs of the pigmy possum have recently been recognized by the extension of the Bogong National Park to provide the necessary sub-alpine environment; it is known that the ground parrot requires regular burning of its habitat to ensure adequate food supplies, and so on. But, as yet, it does not appear to have been recognized that the lyre-bird, also, needs a 'managed' environment. The advent of white settlement in Australia has imposed all sorts of constraints on the forests and native fauna, which are unable to defend themselves against the relentless forces of destruction. The message is clear: unless we take steps to understand the needs of our natural environments, the insidious pressures of 'progress' and time will eliminate them not only from places like Sherbrooke Forest but also from our national parks and other conservation areas as well. Every area of land needs to be managed (and this may entail regular burning) in order that it may continue to provide a suitable habitat for the special flora and fauna which distinguish that particular area of land from others.

Somehow we must strive ever harder to impress the needs of our natural heritage on the land-managing agencies and on governments, because they control our manpower and our money. There is not much time left: will the lessons to be learnt from Sherbrooke Forest go unheeded? Have the lyrebirds of Sherbrooke been sacrificed in vain? I hope not.

My own association with the lyrebird spans over fifty years and continues. In earlier works I endeavoured to share my experiences with others, but, lacking adequate knowledge, I erred in some ways. The purpose of this book is to express my appreciation of the efforts of those pioneers and others who have contributed to the fund of knowledge of the lyrebird, as well as to describe some of my own experiences with that intriguing subject, while at the same time making amends for past errors.

This book is a completely new work, although inevitably some of the material from my earlier publications has found a place. I hope that the inclusion of an account of more recent work on the development of the lyrebird's tail, an analysis of the lyrebird's song, and a discussion on the role of the endocrine system in the life of the lyrebird (conjectural though it may be), will prove interesting to the general reader as well as to those of scientific bent. I have taken the opportunity to tell the story of lyrebirds in Tasmania (surely a conservation 'first', however controversial it may be in the eyes of some), which throws valuable light on the management of natural areas to promote the well-being of the lyrebird in other parts of its range. I have also included an account of a phenomenon never previously observed: a male lyrebird actually rearing a motherless chick. This remarkable behaviour took place at Healesville Sanctuary in 1985.

Despite the advances already made in elucidating the life-story of the lyrebird, it is clear that much more remains to be done. It is certain that greater efforts to extend our knowledge of Australia's outstanding bird would be well rewarded.

L. H. Smith

# Acknowledgements

I am indebted to many people for their help in the preparation of this book; but, in attempting to list those who have assisted in various ways, there is always a danger of omitting the name of a valued helper. I hope therefore that all concerned will understand that I am indeed grateful for such assistance. I am, however, particularly indebted to the following: all those pioneers who have shared the fruits of their labour with me through their numerous publications, along with the organizations and editors who have made such publications possible; all those editors who have found space in their journals and magazines for my own narratives, and those who have written to me to share their personal experiences, thereby enriching my knowledge and appreciation of our common interests; the staff of the State Library of Victoria and the La Trobe Library, and of the Mitchell Library in Sydney, for their courteous help in locating the many publications which I consulted in the course of this work; to Alan McEvey and Belinda Gillies, for facilitating access to the lyrebird collections at the National Museum of Victoria and for the trouble they took, in collaboration with Walter Boles of the Australian Museum in Sydney, to make it possible for me to examine the Australian Museum collection of lyrebirds; Drs David Snow and Ian Galbraith for permitting me to examine lyrebird skins at the British Museum of Natural History; Dr Dean Amadon and his colleagues, for introducing me to the lyrebird collections at the American Museum of Natural History, New York; the Sherbrooke Survey Group, for providing me with identifiable subjects in Sherbrooke Forest, thereby facilitating my studies on the development of the tail plumage of the superb lyrebird; Dawne McPherson for her invaluable help in field work in Sherbrooke Forest; the several proprietors of Sherbrooke Lodge and especially my good friends Thelma and Harry Wileman, for hospitality over the years; my good friend Syd Curtis, whose generous gift of tape recordings of the superb and Albert lyrebirds in Queensland helped me to acquire a better understanding of these birds and who assisted in identifying items of mimicry in their songs; Jean Harslett and Bill Goebel for information regarding the superb lyrebirds of Girraween; Norman Robinson, who has given helpful advice and broadened my knowledge of the subject by his generous gift of tapes of the songs of lyrebirds and other birds; Mr B.H. Pratt, Director of Conservation and

Agriculture, ACT, and the staff of the Tidbinbilla Nature Reserve, for facilitating my acquaintance with the lyrebirds of that region.

I am grateful to the officers of the Sir Colin MacKenzie Zoological Park at Healesville, Victoria, especially Neil Morley and Kevin Mason, for the loan of lyrebird tail feathers and for helpful discussions on various aspects of lyrebird behaviour.

Our Tasmanian experience would have remained a dream had it not been for the enthusiastic co-operation of my old friend Peter Murrell, Director of National Parks and Wildlife in Tasmania, and his staff, and of Trevor Westron and his colleagues at Mount Field National Park, and Andrew Skinner and his staff at Hastings Caves Scenic Reserve. I thank Gerrard Cross and his staff, of Australian Newsprint Mills at Maydena, Tasmania, for introducing me to the lyrebirds of the Florentine Valley.

Despite the excellent library facilities available in Australia, it is sometimes difficult to obtain certain papers and I am especially grateful to the following for providing me with copies of papers not available locally: Professor Hewson Swift of the University of Chicago, for copies of papers by Drs F.H. Lillie and Mary Juhn; the Secretary of the Linnean Society of London, for a copy of Davies's first scientific paper of the lyrebird; Dr Wesley E. Lanyon of the American Museum of Natural History, for copies of several papers; Roger Cressey of the National Museum of Natural History, Washington, DC, for copies of papers by John E. Ward; and to Dr Storrs Olsen of the Smithsonian Institution.

I gratefully acknowledge the assistance and encouragement of the late Professor Carl Welty of Beloit, Wisconsin; Dr Peter Stettenheim of New Hampshire; and Professor Alan H. Brush of Storrs University, Connecticut, whose patient efforts have helped me to a better understanding of the subject.

I owe a special debt of gratitude to my wife Margaret, who has shared the joys of adventurous days in the forest and the rigours of winter camp life in many parts of the lyrebird's haunts and who has carried an undue share of the domestic burden while I was pursuing the lyrebird with camera and tape-recorder. Without that unfailing support over more than forty years, this book could never have been written.

Words alone cannot express my gratitude to my old friend Stanley S. Payne for his encouragement and interest in my work and to the Edward A. Green Charitable Foundation, of which Mr Payne is the trustee, whose generous sponsorship has enabled the price of the book to be held at a reasonable level.

I gratefully acknowledge the help of Cathryn Game for her patient and skilful editing of the manuscript.

In conclusion, I am grateful to my publishers for the opportunity of sharing my experiences and thoughts with others and for adding my pleas to those of a growing band of people who are concerned for the well-being of our natural heritage.

L.H. Smith

# 1 | Historical introduction

## Discovery of the lyrebird

One of the key figures in the story of the discovery of the lyrebird was an ex-convict, John Wilson, who had spent several years with the Aborigines and gained considerable knowledge of the bush. However, even when the colony was only ten years old, the administration was becoming embarrassed by troubles arising from the association of a number of ex-convicts with the natives. Governor John Hunter issued an edict requiring all white men who were living with the Aborigines to deliver themselves to the nearest peace officer or face severe penalties. Wilson was among those who yielded to the Governor's persuasions, and was pardoned.

Governor Hunter became aware of

Wilson's claims to have seen a strange 'pheasant-like' bird while living with the natives and, because of his natural interest in birds, Hunter's curiosity was aroused. But the Governor also had less pleasant matters on his mind: among those enjoying the hospitality of the government were several Irish prisoners who had become consumed with the notion that somewhere to the south-west of Sydney existed a Shangri-la peopled by white men who had cast off the harsh life under the government, and longed to join this select band. They spoke incessantly of wanting to spend the rest of their days in this Utopia.

Governor Hunter saw an opportunity of killing two birds with the one stone. He organized an expedition to seek the Promised Land and to obtain information about native fauna. The exploring party was to be led by John Wilson and included four Irish prisoners and two civilians. However, when Hunter learned that the recalcitrant Irishmen planned to kill their guards and seize the supplies, he had the leaders of the plot flogged and strengthened the party by the addition of four soldiers.

The party left Parramatta on 14 January 1798 and headed south-west but, after ten days, the rough terrain proved too much for the prisoners and they were glad to return to Sydney along with their guards. The three civilians pressed on with their journey of discovery. They did not, of course, find a new world peopled by white men, but they did discover some very fertile tracts of land which Governor Hunter had further investigated by a second exploring party, led by Wilson shortly after the first party returned to Sydney. These areas, in the Southern Highlands of New South Wales, were later developed to form the basis of such places as Mittagong, Bowral, Moss Vale, Berrima and Goulburn.

However, our interest is in the first exploring party and especially its youngest member, 22-year-old John Price, who had been in service with the Governor at his home in Sydney. Because John Price could read and write he was able to keep a diary which facilitated the second journey. Price recorded that

on 26 January 1798, ten years after Governor Phillip landed with the First Fleet at Sydney Cove, he shot a bird of 'pheasant-like appearance'. Although Wilson had brought the first news of the bird in 1797, the honour of collecting the first specimen belongs to John Price. By the time he and his companions arrived back at Sydney they had walked some 385 km and, although they saw several more 'pheasants' on the trip, this was the only one collected and duly presented to Governor Hunter.

Surely this was a remarkable feat for anyone, let alone a young man inexperienced in the ways of the bush, and indeed it seems that he and his companion owe their survival and safe return to the resourceful Wilson who had learned so much of his bushcraft from the Aborigines.

In his diary John Price recorded that on 26 January 1798 he shot 'a wild bird about the size of a pheasant, but the tail of it very much resembles a peacock with two large long feathers which are white, orange and lead colour and black at the ends; its body betwixt a brown and green; brown under the neck and black upon his head. Black legs and very long claws.'

Wilson collected two additional specimens on the second journey (of about 560 km), making Governor Hunter the proud custodian of three specimens of the new 'native pheasant', not yet officially named.

Towards the end of 1798 the Governor sent one specimen to each of three friends in London. One of them went to Sir Joseph Banks who in 1770 had accompanied Captain James Cook on his voyage in the *Endeavour* and had, to his credit, demonstrated an interest in the flora and fauna of the new colony. The second skin was sent to Mr Arthur Harrison, while the third was given to Lady Mary Howe, a daughter of Admiral Howe who had fought against the colonists in the American War of Independence and later against the French. It is not known who received which particular specimen, so it remains uncertain whether Lady Mary Howe received the original bird taken by John Price.

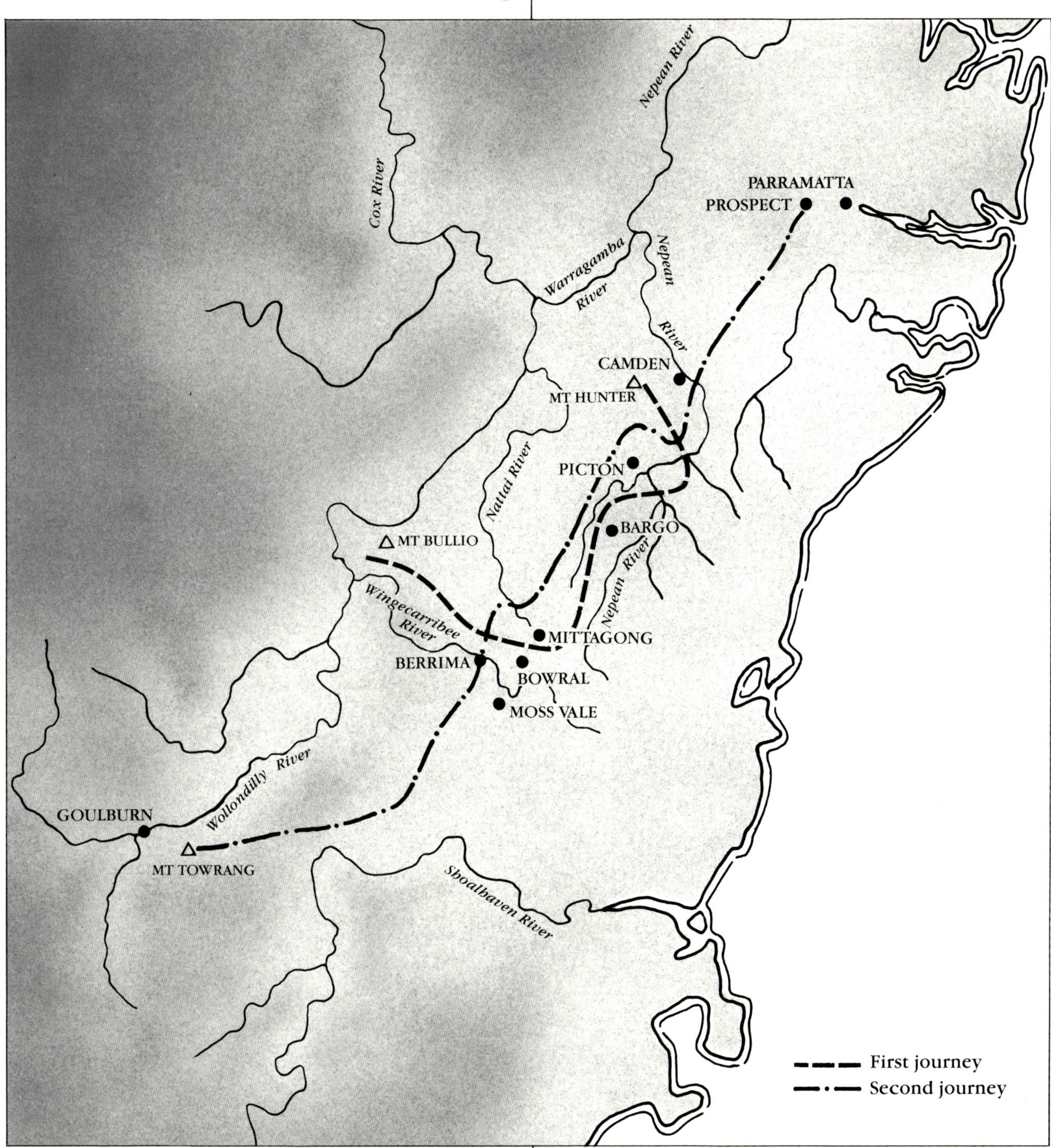


**Map 1** *Lyrebirdmen's journeys south-west of Parramatta, 1798. From The Romance of the Lyrebird, A.H. Chisholm, 1960, by courtesy of Angus & Robertson.*

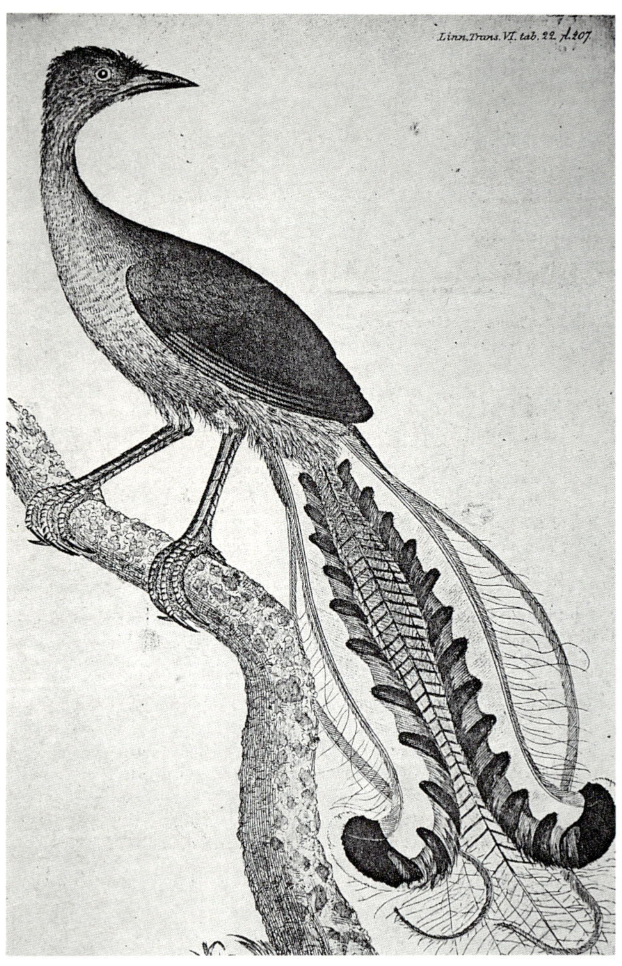



*A copy of the first painting of the superb lyrebird ever made, by Major-General Thomas Davies, in 1800. By courtesy of the Linnean Society of London.*

# The first scientific paper

It is remarkable how many important events are brought about by sheer chance. So it was when Major-General Thomas Davies called on Lady Mary Howe one afternoon in the summer of 1799 and learned that she had in her possession the skin of a strange bird recently received from New South Wales. Davies, a keen bird-man, immediately recognized that Opportunity was knocking on his door and set about describing the new bird in a paper which he presented to the Linnean Society of London on 4 November 1800. He named the new species 'Menura superba', meaning 'the superb bird with the crescent-moon tail'. So the Major-General, who had never seen a live lyrebird or heard its song, became the first to publish a scientific account of the famous lyrebird. Davies's paper was not published until 1802.

Davies mentioned in his paper that he had also seen two other specimens of the new bird in the possession of Sir Joseph Banks, but did not comment further. He illustrated his paper with a colour print made from a painting done by himself. Because of a delay which occurred in the publication of his paper, Davies was able to add a note describing two additional specimens, one of each sex, which Banks had recently received from Governor King (who had succeeded Hunter towards the end of 1800). He does not comment on the male but, of the female, he says that 'with little deviation, the same characters and colours will serve for both of them'. The female was described as having 'a little rufous under the throat'.

The delay in the publication of Davies's paper proved advantageous to Lt. Col. David Collins who, having accompanied Governor Phillip in 1787–88 and remained in the colony for eight years before returning to England, was able to use the name *'Menura superba'* in his description of the bird in his *Account of the English Colony of New South Wales* which appeared towards the end of

1802. He was also able to include some notes he had received from 'persons resident in the country and who had been eye-witness to what is here told' on the lyrebirds' preferred habitat, feeding habits and song, including its remarkable powers of mimicry. Collins illustrated his account with a colour print of a painting by a 'competent artist' of a mounted specimen in the possession of Arthur Harrison, who also had a specimen of a female lyrebird.

The liberties which artists took with their subjects are well illustrated in some early works on view in the Mitchell Gallery in Sydney; there is even one depicting a lyrebird with sixteen filamentaries. A further striking example of artistic licence is a fine engraving by the artist Giacomelli which appeared in a fascinating book, *The Bird World*, by W.H. Davenport Adams (1880). The author describes it as portraying the lyrebird in 'admirable fidelity to nature'. Among the artists interested in Australian birds was Polish-born Gracius Joseph Broinowski, whose *Birds of Australia* was published in six volumes in 1891 (Mathews, 1942). Broinowski's painting of the superb lyrebird is reproduced in this book.

As might be expected, once the bird had been brought to the attention of Governor Hunter and his colleagues specimens were sent to England and Europe. While the first lyrebird hunt was in progress, George Bass was making history elsewhere by discovering the strait which now bears his name; but, in the course of his land journeys, he collected a lyrebird which he duly sent to Sir Joseph Banks.

It is clear that the lyrebird in death was moving in exalted company, far removed from any it had known in its native land.

# Finding a name

A variety of names was proposed by the many authors who sought to involve themselves in the lyrebird's naming, and some brief explanation seems necessary. What follows is a resumé of the story.

A copy of an engraving by Giacomelli, in 1880, depicting the superb lyrebird in 'admirable fidelity to Nature'. By courtesy of Thomas Nelson & Sons, London.

The delay in the publication of Davies's paper proved costly to him because, before this had occurred, in 1801 John Latham, a medical doctor with an interest in ornithology, published an Index to Birds and used the name *Menura novae-hollandiae*, New Holland Menura. Had Latham adopted the name first proposed by Davies, the bird would have been known as *Menura superba*. But, by the rules of zoological nomenclature, the name used by Latham has priority. Nevertheless, the original name continued to be used for many years and in his works John Gould referred to *Menura superba*.

As interest in the new bird grew and more specimens, including a number from the Port Phillip District, found their way to England, they came under the scrutiny of Gould who in 1862 wrote:

'Those ornithologists who have examined specimens of the Menura from the Melbourne district have noticed a great difference in the structure of their tails from this lyre-shaped organ in examples from New South Wales. Although on slender grounds, I admit, I have been induced to consider the Port Phillip bird to be a distinct species; I say slender grounds, because I have not seen a sufficient number of specimens from that locality to enable me to say positively that it is really different...The chief difference of the bird I have named *Menura victoriae* is the diminished length of its outer tail-feathers, and their much stronger markings.'

A.J. Campbell (1899) used the name *Menura superba* for the lyrebird of New South Wales, but expressed reservations about the name *Menura victoriae* for the Victorian lyrebird.

In 1912 Gregory Mathews took the bold step of reducing the Victorian lyrebird to the rank of a sub-species only and, in 1916, 'distinguished the form inhabiting southern New South Wales as an intermediate race, having a shorter tail than the typical form, but not so pronounced as in the Victorian sub-species'. He went on to say that 'three sub-species may be separated, viz:

*Menura novaehollandiae novaehollandiae*
Latham, North N.S. Wales

*Menura novaehollandiae intermedia*
Mathews, Southern N.S. Wales

*Menura novaehollandiae victoriae*
Gould, Victoria.'

The confusion over the naming was aggravated when, in 1921, A.H. Chisholm proposed the name *Menura novae-hollandiae edwardi*, or 'Prince Edward's lyrebird' for a very interesting form of the lyrebird which had been the subject of some study by Dr Spencer Roberts (1922) in the 'Granite Belt' of south-eastern Queensland. When discussing this matter in his *Romance of the Lyrebird* (1966), Chisholm seemed to have lost some of his youthful enthusiasm for the exalted status of this lyrebird and recognized it as 'a southern lyrebird in a northern setting', just as Queen Victoria's lyrebird *Menura n.h. victoriae*, has come to be recognized as 'the southern outpost' of the original lyrebird.

There are undoubted differences between the lyrebirds of the northern, intermediate and southern parts of the range. There are sometimes conspicuous differences in the size, shape and pigmentation of the tail feathers and there are striking differences in the songs of lyrebirds from different regions within the range. But it has also been observed that, even in a small part of Sherbrooke Forest, the lyrebirds exhibit striking differences in regard to the size, shape and pigmentation of their tail feathers and sometimes in their vocalizations. In the absence of sound taxonomic reasons for doing otherwise, there appears to be a general concensus among ornithologists that the correct designation of the bird we are considering is 'superb lyrebird *Menura novae-hollandiae* Latham'. While there may be some sympathy for Davies, the prize remains with Latham, who was 'first off the press'. In this book the name 'superb lyrebird' will be used when referring to *Menura novae-hollandiae* Latham.

The vernacular name 'lyrebird' did not come into general use until about a quarter of a century after Davies first described the new bird but eventually it superseded such names as 'pheasant' and its many variants.

*Copy of Collins's colour print of a painting by a 'competent artist'. From the Facsimile Edition of* An Account of the English Colony, New South Wales *(1971), by courtesy of the Libraries Board of South Australia.*

*G.J. Broinowski's plate of the superb lyrebird from his* Birds of Australia *(1891), by courtesy of the Trustees of the State Library of Victoria.*

*A mature male superb lyrebird singing on a large log makes an easy target for a sportsman's gun—or a camera.*

*The gray (or crested) gallito* (Rhinocryptid lanceolata), *suggested as distant relative of the lyrebird. From* Birds of the World, *courtesy of Western Publishing Co. Inc.*

*John Gould's painting of the superb lyrebird, here reproduced by courtesy of the State Library of Victoria.*

# John Gould joins the quest

Of course, the lyrebird was not the only species of which specimens were sent to England and Europe, to scientific institutions and museums as well as to private collectors who were eager to exploit the bird-life of the new colony. John James Audubon, the indefatigable American naturalist and artist, was already active in the field before John Gould was born in 1804, and it is possible that Audubon's passion for the birds and animals of America and France influenced the youthful Gould. At all events, at an early age John Gould embarked on a career which was to make his name honoured throughout the world. In 1837 he began to publish his *Birds of Australia*, but found that he had insufficient knowledge of the subjects he wished to describe. He therefore decided to visit Australia, taking with him his wife, who was an accomplished artist and was to paint some six hundred of her husband's birds, and their eldest son. The Goulds left England in May 1838 and returned in August 1840.

After spending some time in Tasmania (where they first landed), they went to South Australia, then back to Tasmania before going to New South Wales where, among other things, Gould sought to improve his knowledge of the lyrebird. He was fortunate in having a number of valued assistants and, at the end of his visit, the number of known Australian birds had grown from three hundred to six hundred.

Gould spent much time in his quest of the lyrebird, mainly in the Illawarra district south-west of Sydney, but also worked in the Liverpool Ranges. He frequently heard the lyrebirds singing and obtained many specimens but did not see a male bird displaying, nor did he find a nest occupied by a chick. This was because he was in the lyrebird's domain at the wrong time of the year. In his *Birds of Australia* Gould wrote feelingly of the frustrations to be endured in stalking lyrebirds:

'Of all the birds I have ever met with, the Menura is by far the most difficult to procure. While among the brushes I have been surrounded by these birds, pouring forth their loud and liquid calls, for days together, without being able to get a sight of them; and it was only by the most determined perseverance that I was enabled to effect that desirable object, which was rendered the more difficult by their often frequenting the almost inaccessible and precipitous sides of the gullies and ravines covered with tangled masses of creepers and umbrageous trees.'

However, despite the difficulties, Gould was able to make a number of interesting observations on lyrebirds; he remarked on their habit of chasing one another through the forest as if in play and on

'the curious habit they have of forming numerous small hillocks which they constantly visited during the day and upon which the male is continually trampling, at the same time erecting and spreading out his tail in the most graceful manner, and uttering his various cries, sometimes pouring forth his natural notes, at others mocking those of other birds, and even the howling of the Dingo.'

# Early Victorian lyrebird men

The first account of the lyrebird in Victoria was given in 1846 by G.H. Haydon in his informative narrative, *Five Years in Australia Felix*. Following the discovery of Gippsland by Count Strzlecki in 1840, Haydon led a survey party consisting of twelve white men and several black police, with the object of opening up a route connecting the Port Phillip and Westernport districts with Gippsland. This was rough country; it almost claimed the lives of Strzlecki and his party who spent twenty-six days in a vain attempt to cut a track; but, no doubt aided by the knowledge acquired by Strzlecki, Haydon accomplished the task in thirty-five days.

Haydon's party left Westernport on 23 April 1844 and proceeded eastward; early on the morning of 3 May Haydon was awakened by the singing of several 'pheasants' 'imitating the sounds one heard in the bush to great perfection'. Intent on providing himself with fresh meat for breakfast, he proceeded, gun in hand, towards the singing bird.

> ' The sun having risen [he wrote] induced it to commence its morning song, but the natural note (blec blec) was almost lost in the multitude of sounds it was producing – the croak of the crow, the scream of the cockatoo, the doleful cry of the more-pork, the chattering of the parrots, the ridiculous hooting of the laughing jackass, the howl of the wild dog, all made in quick succession. '

The bird was on its mound; he stalked it and was about to fire, when a gun went off nearby and the bird's head was blown off by one of the black police who had had the same urge as Haydon. The latter, however, claimed the prize and 'regaled' himself. Haydon described the female as 'a very unattractive bird, having but a poor tail, nothing like the male'. He added that the female laid only one egg (each season) of a 'slate colour with black spots', the nest being about three feet in circumference.

Although Governor Hunter had sent Sir Joseph Banks a specimen of the 'Bird of Paradise' (to use another name bestowed on the lyrebird), along with an egg, the latter appears to have been damaged. The first occasion when the egg of the superb lyrebird was publicly displayed was at the Melbourne Exhibition of 1854, when F.J. Williams and J. Leadbeater exhibited specimens of the lyrebird along with the egg.

Among those to study the lyrebird was Dr Ludwig Becker, who came to Australia by way of South America, in 1850. Ludwig Becker was a man of many parts — an artist, a scientist and a geological surveyor. He was appointed artist, naturalist and geologist to the Burke and Wills expedition on which he perished. (His journals and sketchbook were published in 1980.) He is also remembered for having designed the emblem of the University of Melbourne, for having discovered a vivipar-

ous fish and the sea-horse in Hobson's Bay and for having inadvertently played a part in the hatching of the first lyrebird chick in captivity.

With the permission of Williams and Leadbeater, Becker made a pencil drawing of the lyrebird egg which had been displayed at the Exhibition and sent it to Professor Kaup in Germany; the latter passed it on to Professor Cabanis in Berlin for publication. Thus did the lyrebird egg first become known to Europe.

The enterprising Becker persuaded a native named Simon and one of his relatives to obtain a lyrebird's nest and egg for him. Simon was the son of one of the chieftans of the Yarra Yarra tribe who had concluded the sale of land to John Batman in 1834. A nest and egg having been found on 31 August 1854, the two natives arrived back in Melbourne on 4 September. During those five days' travel, the egg had been wrapped in the folds of Simon's possum-skin coat. He did not know it then, but in fact he was acting as a human incubator and, shortly after he had handed the egg to Becker, they had the remarkable experience of seeing a lyrebird chick break out of the shell. Becker was thus enabled to describe the new-born chick for the first time (1855). He rewarded the natives by presenting them with sketches of Simon and his family...if only those sketches could be found today, what a price they would bring!

# Exploitation

The lyrebird soon became a prize for the sportsman's gun and a source of revenue for traders who specialized in procuring specimens of Australian birds and animals. Among the sportsmen to visit Australia was J. D'Ewes who, in his informative narrative, *Shooting in Both Hemispheres*, gives an account of his brief stay in Melbourne in the winter of 1853 before taking up his appointment as a police magistrate at the Ballarat goldfields.

While in Melbourne, then a 'city of tents', he made the chance acquaintance of a doctor who organized a shooting excursion to the Dandenong Ranges. After some senseless shooting of 'twenty couples of snipe', some plovers and a black swan, the party began an abortive search for kangaroos and, next day, D'Ewes began his quest of the lyrebird. From his place of concealment, he soon heard 'those very mocking notes' of his invisible quarry. But, shortly, he perceived a movement in the long grass and, in his eagerness, fired both barrels, only to have the mortification of seeing 'a magnificent cock bird, with tail erect and lyre that would have put Apollo's in the shade emerge from the tufted grass and dart into the jungle'. This lyrebird escaped, but so many thousands did not.

Among the professional collectors was an American, Sherman F. Denton, who, in his frank and readable book (1887), tells of the collecting of the lyrebird and other birds in the Plenty Ranges (better known now as the Kinglake district). He and his companion were entranced by 'the most exquisite mimicry of many of the songsters of the wood, varied by sounds resembling the clear tones of a distant bell, the rattle of a rickety wagon, raspings and gratings that made a cold chill run down one's back, whispers, moans, cries and laughter'. 'Sometimes the song, with a volume like a large organ, was loud and sweet; there was a mellow richness that reminded me of the clarinet,' wrote Denton.

Denton tells how the birds were shot as they ran along large logs to escape. At Bald Hill (in what is now Kinglake National Park), Denton stalked a lyrebird which eventually went to a mound and

> ' strutted about, spreading his magnificent tail, and making the woods echo with his wonderful song. Like a fox stalking a hare, I crawled through the ferns toward him and when near enough sprang up and instantly fired. When I came to the spot he lay stretched upon the mound, the tail spread, and every feather as perfect as when alive. '

It was not until 1879 that any sort of legislation was enacted to protect the lyrebird, but even that was practically useless, because the lyrebird was classified as 'native game' and

protected only for a short period during the breeding season of other birds, that is, during the spring, which is not the breeding season of the lyrebird.

Under the circumstances, therefore, it is not surprising that men of 'enterprise' sought to make easy money by shooting lyrebirds and selling their tails. In an informative little book entitled *A Sketch of Natural History in Australia,* Frederick George Aflalo (1896) relates how

> ' two enterprising brothers recently employed a number of men to shoot the luckless birds in which, after some practice, they were unfortunately so successful that 500 dozen of the beautiful tails were reported to have reached Sydney in the course of a few weeks. It is not difficult to understand how at this rate the price of tails which, according to Bennett, was as high as 30/- a pair fifty years ago, should have fallen to one third of the price, at which figure I could have bought a hundred pairs in Sydney had I been so minded. '

In his classic work, *Nests and Eggs of Australian Birds*, A.J. Campbell (1900) provides a lively account of his experiences in Gippsland. Let him give his own account of his first excursion in the Neerim district, towards the end of the summers of 1875 and 1877. Referring to the grandeur of the forests of those days, he wrote,

> ' The reader may gather some idea of the semi-tropical growth, so to speak, if he can imagine three great forests rolled into one:— firstly, thickly studded elegant fern-trees entwined with various parasitical creepers, forming fairy-like bowers carpeted with a ground scrub of innumerable ferns; secondly, trees of medium height, such as sassafras, musk, pittosporum, native hazel, blackwood, and other acacias, etc., and, thirdly, towering above all, a great forest of gigantic eucalypts. Within and under the triple shades of these leafy solitudes, is the true home of the wonderful Menura. '

During the two trips Campbell was able to shoot ten male lyrebirds and as many females as he required for his collection. Let us in our imagination accompany him as he tells of his experiences:

'You patrol leisurely up a gully or along the survey lines till you hear a bird merrily whistling on its hillock, or dancing mound, a little distance in; then you commence carefully — oh, so carefully, for one false step, an extra shuffle of the leaves or the snapping of a twig underfoot, and your prey simply disappears as if by magic — to crawl on your hands and knees, as often as not wriggling snake fashion on your stomach through ferns and scrub, from stump to stump and from tree to tree. Listen! the bird stops singing as if instinctively knowing danger is approaching, whereupon you have to become like a statue, fixed to some fern root, and dare not move a muscle even if you feel a land leech attacking your legs, or a large mosquito attacking the tip of your nose. Presently the bird commences whistling as merrily as ever. On you creep, every yard nearer, so that with the excitement your heart increases in palpitation till it throbs so loudly that you fancy the bird will hear it. All the while, the close humid scrub bathes you in perspiration, while great beads stand upon your forehead; then, rolling off, patter on the dried leaves beneath you. Affairs are desperate now, for at last you are within shooting distance and are peering through the ferns with uplifted gun and finger trembling on the trigger; but, alas, the bird possessing sharper eyes than you discovers you first, and is that very instant off noiselessly and unperceived. There is no alternative left but to retrace your steps to the track and your chagrin can be better imagined than expressed. This operation you may repeat on an average five times before you get even the slightest possible chance of shooting a bird. But I found the females easy to bag, for they frequently leapt into the trees overhead to survey me.'

However much we may deplore his actions in shooting lyrebirds and taking eggs from nests, we may nevertheless feel envious of people like Campbell who, in those more leisurely days, found time to savour the pleasures of the bush. It did not matter that the track was slippery or non-existent; the forest was like a mighty temple whose incense was the aromatic fragrance of the sassafras, dogwood, musk, hazel and eucalypts, especially after rain.

I recall a frosty morning in June, when I stalked a lyrebird in Gippsland, near Balook, with my tape-recorder, when every cautious step up that hill sent a sharp stabbing pain through my chest, while my wife, from the comfort of her warm bed, through the open front of our tent, watched the lyrebird on his lofty perch in a tall blackwood tree, as he poured forth his greeting to the dawn. This is a place where, on a black night, the stars shine ever more brightly and the air is freezing, and the stillness of the night is broken only by the hoot of the boo-book or the yelp of a fox, which sends a cold shiver down the spine …and I yearn to return there, just once more.

I have not spared the morbid details of the shooting of the lyrebird, while fully aware that there will be many who share my feelings of revulsion. It is one of the major tragedies of history that, wherever white people have gone in their quest for world domination, they have plundered the newly discovered lands and systematically destroyed the wildlife in the most wanton manner. Nothing can ever excuse this ruthless destruction of our natural heritage and there is an urgent need to protect our remaining birds and animals along with their natural habitats, against the avalanche of 'progress' and development which threatens to envelop them. We are still a young country, yet much of our natural heritage lies buried beneath the waves of unplanned land development, and governments must be encouraged to continue their conservation activities to ensure that past mistakes are avoided.

# Sherbrooke pioneers

Gradually, the wholesale shooting eased and naturalists began to arouse public interest in the lyrebird. Among those who helped in this work was Tom Tregellas, who died on 10 October 1938 at the age of 74 years. Several years before this, in the mid-1920s, he had set up a bush home in a large hollow log, on the western slope of Hardy's Gully, in Monbulk State Forest, about 50 km east of Melbourne. He constructed an open-air fireplace along with a large meal table, and soon won the confidence of the lyrebirds nearby. They frequently visited him in his log cabin. He constructed narrow paths through the scrub to gain access to the display mounds and took many photographs, even as far back as 1920. Unfortunately, in those days, neither cameras nor films were capable of coping with the unfavourable conditions prevailing in the forest, especially during the winter. In a report to the *Emu* in 1930, Tregellas remarked somewhat wryly that to take a successful photograph required an exposure of four seconds, 'hence the bird must be very still'.

Tregellas found many nests and studied the behaviour of the female lyrebird; he remarked that one particular female imitated the alarm call of the magpie, the petulant call of the young podargus, the golden whistler, the pilot bird, the currawong, the barking of a fox terrier and the yelp of a fox. From this we learn that dogs and foxes were already in the forest. He banded every one of the many lyrebird chicks which he found in their nests, but was disappointed at not seeing any of them after they had left the nest.

Tom Tregellas and Michael Sharland (1931; 1981) assisted the Australian Broadcasting Commission in organizing the first national broadcast, through 3AR and 3LO, of the song of the lyrebird, direct from Sherbrooke Forest, on 5 July 1931. The broadcast was relayed to all states and was heard with perfect clarity.

The other great pioneer lyrebirdman was Ray T. Littlejohns (1926) who began filming the lyrebird in Sherbrooke Forest in 1925. Ray Littlejohns was not only an ornithologist of high repute but also a photographer of infinite patience and remarkable skill. His written works reveal a sensitive mind and a versatile pen. He used every available avenue to focus public interest on the lyrebird. He was associated with the Australian Broadcasting Commission in organizing radio broadcasts of the lyrebird's song direct from Sherbrooke Forest during the winters of 1932, 1933 and 1934. That part of the forest in which these events took place became known as the 'Broadcast Area' and was a familiar haunt of later generations of lyrebird seekers. It was here that over the years I made the acquaintance of such feathered celebrities as

'Silvertail', 'The Wanderer', 'The Broadcaster' (so named because in 1952 he was recorded by the ABC for broadcasting through Radio Australia), 'The Crescent' and 'Crossed Lyrates'.

I did not meet Ray Littlejohns until April 1939; on that memorable Sunday morning, my wife and I were stalking lyrebirds in an area a little to the north of the Falls, and had attached ourselves to a mature male lyrebird whose mate and chick of the previous season were scratching nearby. I was learning just how clever lyrebirds can be: this one knew exactly when to move on a little, just as I was about to press the shutter release, and how to tantalize me by placing bracken stems or small sticks between himself and my camera. I had been enduring these frustrations for half an hour and was just on the point of taking 'the picture of my life' when a voice from the gloom fell on my startled ears. I turned to greet the intruder; it was Ray Littlejohns. He kindly led me out of the wilderness to the firebreak area and introduced me to 'Timothy', the reigning lyrebird monarch. My association with these two 'greats' of Sherbrooke Forest was to extend over many years and from them I learned much about lyrebirds and the care required to photograph them.

In the years that followed I often met Ray in the forest and we usually had lunch together, sitting on a particular large log which was used as a concert platform successively by Timothy and Spotty, the two birds which dominated the firebreak area in Sherbrooke Forest for more than thirty years.

Ray Littlejohns died in 1961, aged 68, but his memory lives on. His many friends combined to build a stone seat in a part of the forest where Ray was wont to linger as he waited for his favourite bird to sing, and I relive those days when I take my lunch on Ray's memorial seat.

The Sherbrooke Survey Group was established in 1958, under the guiding hand of Ralph Kenyon. I helped to obtain the necessary permission from the Fisheries and Wildlife Department to undertake a banding programme on lyrebirds in Sherbrooke Forest and in

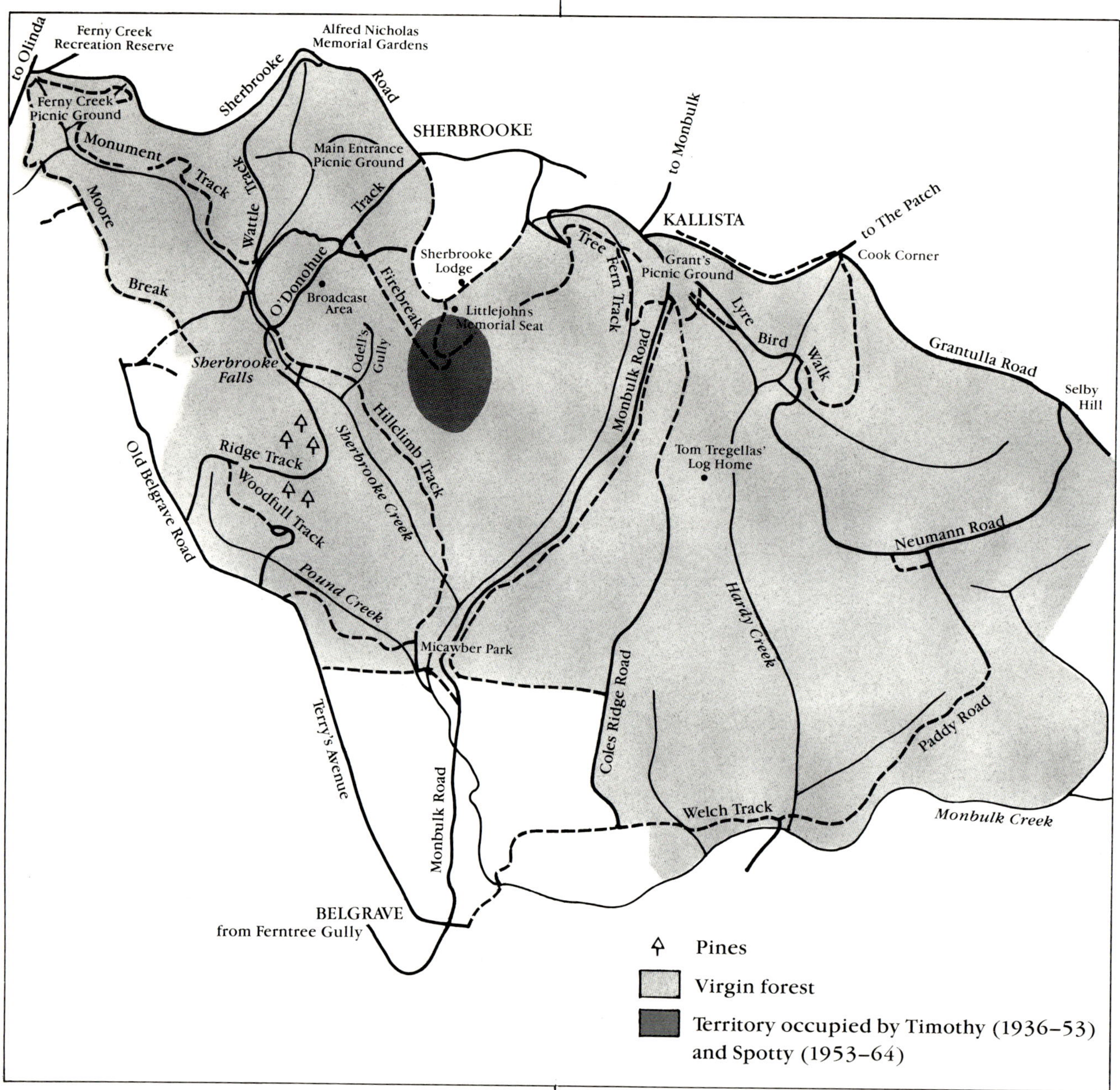


**Map 2** *Map of Sherbrooke Forest, showing principal features referred to in text. Based on a map by the Forests Commission, Victoria.*

detailing the projected research programme. The signatories to the letter of application to the Department were Ralph Kenyon, Ina Watson and myself. Unfortunately, professional duties precluded' my active participation in the fieldwork of the Group, whose patient efforts have resulted in the publication of important scientific papers on different aspects of the life history of the lyrebird.

2

# General observations on the superb lyrebird

The mature male superb lyrebird is widely acclaimed as one of the most interesting members of the avian world. The glittering spectacle of his long tail feathers in display and his loud ringing notes, and his remarkable powers of mimicry, have attracted countless visitors to his haunts from all over the world. I have seen young children standing or squatting on the ground as, in silent wonder, they watched a lyrebird scratching industriously for food, while a yellow robin or a pilot bird hopped fearlessly in and out of his worm patch, gathering titbits. In 1939 I saw a Japanese diplomat crawling on hands and knees through the damp chilly wintry forest in his efforts to steal a little closer to Sherbrooke's favourite lyrebird of the day. I have seen groups of men and women who had travelled long distances from their metropolitan homes so that they could stand in silence to watch a lyrebird perched on a low branch and hear his enthralling melody. I have guided men and women of science in their eager quest of the master songster and have been moved by the expressions of rapture on their faces when they finally won their first glimpse of a lyrebird displaying on a mound and heard his wondrous song.

Although one early writer expressed the view that the female lyrebird is a very 'unattractive bird', I hope to show in the following pages that she is in fact an extremely interesting creature and that the young lyrebirds also have their special attributes.

# Early origins

Although the designation 'pheasant' persisted for many years after the discovery of the lyrebird, it was early recognized as a separate species (Davies, 1802). Latham, whose name is included in that of the superb species, considered it to be a gallinaceous bird, but John Gould, in 1853, expressed the opinion that the lyrebird was not gallinaceous and placed it among the Passeres or perching birds. However, Bartlett (1867) argued that the lyrebird was related to the birds of paradise. Then in 1875, Professor Alfred Newton, of Cambridge University, proposed that the lyrebird was allied to another Australian form, *Atrichus* (Atrichornis), the scrub-birds, of which there are two species — the noisy scrub-bird (*A. clamosus*) and the rufous scrub-bird (*A. rufescens*). These birds are somewhat smaller than a thrush, while the lyrebird is rather smaller than a pheasant.

But the age-old question remains: did the lyrebirds' ancestors originate in the Northern Hemisphere and reach their present homes by way of South-east Asia or are they derived from some ancient form that originated in the Southern Hemisphere? A casual glance at some of the birds of paradise so magnificently illustrated in Iredale's book of that name (1950) might well suggest a possible relationship between these birds and the lyrebirds. The long fine wire-like central feathers and the 'filamented' structure of the main rectrices of the tail plumage of such birds as Blood's bird of paradise, the greater bird of paradise, the red-plumed bird of paradise and Gouldie's bird of paradise, to mention but a few, show some resemblance to the corresponding tail feathers of the superb lyrebird. Indeed, Sibley (1974), on the basis of

comparative studies on egg-white proteins, suggested that the lyrebirds might be related to the birds of paradise and the bowerbirds, but Feduccia and Olson (1982), on the basis of comparative osteology, concluded that the presumed relationship between the lyrebirds and the birds of paradise and the bowerbirds is untenable. They suggested a possible relationship between the lyrebirds and the tapaculos (Rhinocryptidae), a family of ground-living birds which inhabit the dry grassy plains and scrublands of the thick undergrowth of the montane forests of South America.

They also remarked on the fact that, in 1841, Eyton had drawn attention to the resemblance between the Menuridae and the South American genus *Pteroptochos,* another member of the Rhinocryptid family and that, in 1859, Cabanis and Heine had once included the Menura in their family Pteroptochidae (= Rhinocryptidae).

The tapaculos, which range in length from 11 to 25 cm, are most numerous in the temperate third of South America; they have short rounded wings and are poor fliers, but can run and hop swiftly, holding their tails 'more than erect'. One species, the gray gallito (*R. lanceolata*), builds a bulky nest of grasses and bark, with a side entrance. These little birds fascinated Charles Darwin when he was in the High Cordilleras in the Andes, near Valparaiso, in the course of his voyage in the *Beagle* in 1834.

Feduccia and Olson argued that the lyrebirds and the scrub-birds and the Rhinocryptidae originated in the Southern Hemisphere and dispersed before the vast land masses comprising Gondwanaland became separated.

Serventy (1973) had earlier proposed that the lyrebirds and scrub-birds were of southern origin. In the historic past Tasmania and the mainland of Australia were joined by a land bridge. Ridpath and Moreau (1966) adduce evidence to show that about 100 million years ago the western part of Tasmania supported vast stands of Antarctic beech (Nothofagus); but, during the Pleistocene glaciation period, the Nothofagus retreated northward and Robinson (1977) has argued persuasively that the ancestors of the Menurae were likewise forced to migrate northward across the land bridge (which came and went several times during the Ice Ages); but, when the last Great Ice Age came to an end about 12,000 years ago, Tasmania and Wilsons Promontory were isolated from the mainland. The lyrebirds became established in the favourable environment of the Eastern Highlands of the mainland, but were unable to cross the sandy isthmus which joined Wilsons Promontory to the mainland about 5000–6000 years ago. The lyrebird, with its poor powers of flight and its dependence on a wet sclerophyll type of habitat to provide its essentially protein diet, has likewise been unable to cross the drier parts of central Victoria to reach the Otway Ranges in south-western Victoria, despite what would appear to be the favourable environment of that region.

However, Sibley and Ahlquist (1985) have recently concluded that the birds of paradise and bowerbirds are not related and that the latter comprise the sister group of the lyrebirds and scrub-birds. They refute the suggestion that the lyrebirds are related to the Rhinocryptidae. So the quest for the golden thread goes on.

# Range and habitat

The range of the superb lyrebird extends from Ferntree Gully in the Dandenong Ranges, about 30 km east of Melbourne, eastward beyond the Healesville–Warburton districts through the Eastern Highlands and Gippsland to south-eastern New South Wales, continuing through the Australian Alps and mountain ranges of eastern New South Wales through the New England district to the Queensland border. The most westerly limit of the range in Victoria is the Wandong district, about 60 km almost due north of Melbourne (Eddy and Cusack, 1967). The range extends inland in Victoria as far as the Mount Buffalo district (about 320 km) and in New South Wales as far

as Tidbinbilla (ACT) and the Blue Mountains (about 250 km). Before white settlement, the range would surely have been continuous, but the clearing of land for agricultural and other purposes has greatly reduced the areas available to the lyrebird. Blakers, Davies and Reilly (1984) found that about 90 per cent of the lyrebird population occurs in Victoria and New South Wales.

The nature of the vegetation varies throughout the range, but the wet sclerophyll type of forest seems to be preferred. Generally, there is an upper storey of eucalypts or other dominant species, with a middle storey of acacias, such as silver wattle (*A. dealbata*), blackwood (*A. melanoxylon*) and others, and the slopes and gullies are rich in musk (*Olearia argophylla*), hazel (*Pomaderris apetula*), sassafras (*Atherosperma moschatum*) and a wide range of tree-ferns and smaller varieties of ferns. Because the rainfall varies greatly over the range, the nature and density of the vegetation does likewise and, in certain areas of, for example, the Glenaladale district (Gippsland) and the Blue Mountains, at certain times, the harsh country makes it difficult for the lyrebird to win a living.

# Food

The lyrebird's food consists largely of worms, grubs, beetles, centipedes and an occasional scorpion, which are found in the ground or under logs and rocks which are pulled aside to expose the soft moist earth beneath. Sometimes an unlucky land yabbie falls victim to the sharp eyes of the lyrebird, and it is rather amusing to observe the reaction of the cautious lyrebird as the yabbie waves its claws about in self-defence. A great favourite with the lyrebird is the tiny shrimp-like crustacean (*Talitris* sp.) which abounds under the bark and small logs on the damp forest floor. Any 'hoppers' which escape the lyrebird are eagerly taken by the yellow robin, pilot bird or scrub wren which regularly attend the larger bird. The lyrebird is not selective in what it eats; its natural food is essentially pro-

*Spotty finds a large worm and eyes it cautiously.*

tein and occasionally it finds a large worm, which may be 20–25 cm long and about 6 mm thick. While this is a great bonus to the lyrebird, it nevertheless poses problems but, after cautiously inspecting the slow-moving, almost rigid creature, it proceeds to swallow it. The expressions of discomfiture as the worm wriggles inside are very amusing to watch.

# The lyrebird's year

A brief account of the behaviour of the lyrebird during each of the four seasons will illustrate the range of activities in which the birds are involved in the course of a year; later, some of these will be examined in more detail.

In the spring (September–November), the mature male lyrebird is in tail-moult. The tail feathers are usually shed towards the end of August and a new tail is grown over the next twelve to fourteen weeks. During this period, the mature male is often (but not always) solitary, spending most of his time scratching for food, or perching on a favourite branch or stump, preening or sleeping. However, despite the lack of his tail, he sometimes sings and displays on a mound, on the forest floor, on a stump or on a log. Sometimes he shares a mound with another lyrebird in what is obviously an enjoyable mutual display. During this period, also, he may be seen with his twelve- to thirteen-month-old chick quietly scratching alongside. Whether either bird is conscious of their relationship, I cannot say, but I have seen this phenomenon so often that it seems unlikely to be mere coincidence.

During late spring and summer (October–February), the new tail, now resplendent, is used more effectively in display. Now, too, he is seen in the company of his mate from last season along with her chick and, often, with the one from the year before as well.

In January–March the mature male, like other lyrebirds, undergoes a 'head-and-neck moult' whch occupies about a month. Although the bird looks decidedly scrawny during this period, at the conclusion of the moult, his head and throat are very sleek.

With the advent of autumn (March–May), the tempo of his courtship activities (song and display) increases. The frequency and duration of displays increase towards the latter's peak, leading to mating early in winter (June–August). Displays and singing continue throughout this period, especially during June and July, but the tempo falls as spring and the onset of the tail moult approach. Over a period of about ten days, the tail feathers are shed and new feathers soon appear.

After the female lyrebird matures, she spends most of her time working in various ways, in addition to finding food for herself. In brief, she builds her nest in the autumn or early winter and, after having laid her egg, she is tied to the nest for six to seven weeks while she incubates the egg; then she spends the next six to seven weeks tending the chick in the nest. Thereafter, through the spring and summer, she is in constant attendance on the chick, which does not become fully self-sufficient until the onset of the next breeding season. Then the cycle repeats itself; her domestic duties will be described in greater detail later.

After they become self-supporting, at the age of about nine months, young lyrebirds enjoy a comparatively carefree existence while they learn their respective 'trades'. In effect, they are at school, learning what they need to know from some great teachers and from one another, without the responsibilities of their parents. When mature they inherit the full range of duties, according to their sex, of the adult birds.

# How long do lyrebirds live?

A question often asked is: 'How long do lyrebirds live?' There are few authentic records on the subject, but it is known that 'Timothy', the bird which occupied the fire-break area from the early 1930s to the spring

*Timothy, displaying in the firebreak in Sherbrooke Forest, September 1944.*

of 1953, was 25–26 years old when he died. His successor in the firebreak, 'Spotty', was 21–22 years old at death. The cause of death is not known in either case, but the prevalence of foxes in the area suggests that both birds could well have been victims of this scourge of the forest. Most other lyrebirds in Sherbrooke, where the histories are known, have been much less long-lived. Many young birds are taken before they reach maturity, and it is known that the birds banded R and W/W (known to their followers as 'Single Red' and 'Double White') were about nine years and four months, and seven years and eight months old, respectively, when they disappeared from their familiar haunts. I suspect that they were taken by foxes.

Two fledgelings adopted and reared by Mr and Mrs Jack Coyle of New South Wales were at least 15 years old when they met their deaths by accident.

The evidence from other sources is less conclusive, but the information about the lyrebirds which were kept in the London Zoo suggests that a life-span of thirty-six to forty years might be possible, under favourable conditions. Such conditions no longer exist in the Australian mainland forests, which are virtually overrun by foxes and feral cats.

Williams (1881) stated that he often found as many as forty eggs in the ovaries of female lyrebirds he had shot. This suggests that, in the absence of predators, the birds (females, anyway) were genetically designed for a reasonably long life.

In September 1947, I found the nest of what was to become one of my 'friendly females'. She always built on the northern slope above the Falls Picnic Area in Sherbrooke Forest and, during the breeding seasons of 1947–60, I spent much time with her. Her 1959 chick was the celebrated Red/Blue, which later became a well-known member of the Sherbrooke lyrebird community. Assuming that 1947 was her first nesting season and that she was then in her seventh year it would appear that she was about 19–20 years old (at least) when I lost contact with her. Godfrey (1905) described a free-ranging pet male lyrebird (the celebrated 'Jack') which lived on a farm in Gippsland and died in his twentieth year. Curiously, this is about the same age as that of the two female lyrebirds exhibited in the Adelaide Zoo from 1949 to 1967.

# Lyrebirds and bushfires

Almost every year the forest areas of Victoria are severely damaged by bushfires. Many of these fires are man-made, some deliberately lit, others the result of gross carelessness, and some are the result of natural phenomena such as lightning strikes. The loss of wildlife in such fires is enormous, but some animals do manage to survive. In the disastrous fires which ravaged eastern Victoria in February 1983, the Princes Highway and its verges, in the affected areas, were said to have been choked with birds and animals which had moved out of the forests in the van of the fires, in their efforts to escape.

Some interesting facts on the survival behaviour of lyrebirds come from Mervyn Bill (1942), who tells how a Mr Mitchell saved his own life during the fire which swept through the Erica district in central Gippsland on 5 February 1932. Mitchell was working near the Thomson River at the time and, realizing that it would have been futile to attempt to escape via the forest tracks, decided to remain by the river. The severity of the fire may be gauged from the fact that five of his mates perished in it. When the fire descended, Mitchell took refuge in the river and waited for it to pass. He was, however, not alone; because, from eight o'clock in the morning, three hours before the fire reached him, 'the lyrebirds began to flock from the higher country to take shelter in the river and, moreover, they could not be made to move from the positions taken up immediately on reaching the water'. Bill considered that it was instinct which led the lyrebirds to a place of safety, as much as five or six kilometres from their customary haunts, while the fire lasted. And then, within a few days, when the heat from the burning logs had abated, they returned to their original homes. Indeed, a few days after the fire had subsided, Bill (a Forests Commission surveyor), when riding from Erica to his camp on Talbot Creek, from the volume of sound coming from the charred and blackened areas, judged that there were just as many lyrebirds present after the fire as there had been before.

Another remarkable story is told by Frank Howe (1927). There was a disastrous bushfire in the Warburton district in 1925; but, when he visited a familiar gully in July– August 1926, he was surprised to hear the calls of lyrebirds all over the hillside. There was not a vestige of greenery anywhere and both slopes were bare, except for fallen trees. The musks, hazels, blanket-leaf trees and fern were all charred and black; but the lyrebirds were still there in their old haunts. Moreover, he found eleven nests, certain proof that the female lyrebirds had also survived in good numbers. Before the fire, Howe had discovered a platform nest built on a large rock. The platform was destroyed in the fire, but, in the winter of 1926, he found a nest containing an egg, which had been built by one of Howe's known females, on the same rock. Howe thought that the lyrebirds might have survived by taking refuge in wombat burrows which were numerous in the area.

This is a distinct possibility, because I have often seen a lyrebird (for some reason best known to itself) run down into a wombat's burrow in Sherbrooke Forest. Also, during the bushfires which swept across Victoria and New South Wales in January 1939, some men who took refuge in a mine shaft found themselves sharing it with a number of lyrebirds which had overcome their natural timidity and remained there.

Kitson (1905) tells how a number of lyrebirds which survived the great bushfires of Gippsland in 1898 came out of the burnt scrub and fed among the fowls of the farmhouse near the forest. He makes the interesting observation that in times of necessity the lyrebird could become graminivorous, although normally it lives on various forms of protein.

# Preening and bathing

No doubt because the tail feathers play such an important part in courtship activities (on which the survival of the species depends), the lyrebird spends several hours each day very carefully preening its feathers. All lyrebirds do this — females, young birds and even chicks in the nest. There may be as many as six or eight preening sessions each day. The continued use of favourite perches whilst preening destroys the moss, thereby indicating where the bird may be found at some time during the day.

Lyrebirds also like to bathe and, if a pool of water is available, they will make their way to it, usually once a day and often from a considerable distance. They wade into the water, spread their wings, fluff up their feathers and immerse the whole body several times. They obviously enjoy bathing and occasionally sing in the bath. Even winter temperatures do not cool their ardour.

After leaving the pool, they select a perch and shake themselves vigorously, sending a shower of water into the air, even on to the lens of one's camera. Then each tail feather in turn is drawn through the beak to express more water (as in an old-fashioned clothes wringer) and, when the tip of the feather is released, the feather describes an arc as it returns to its normal position, sending a stream of water drops into the air. Lyrebirds may not know anything about centrifugal action, but they certainly take full advantage of this interesting phenomenon.

After the bird has disposed of its bathwater through these gymnastic exercises, it begins a thorough preening with the beak, with frequent visits to the oil gland at the base of the tail, in order to dry and polish the feathers of the entire body, wings and tail. Even the head feathers are preened, with the aid of the claws.

The entire operation may occupy half an hour, after which the bird resumes its quest for food.

*Spotty, stretching his wings, at the end of a preening session.*

# Lyrebirds at play

John Gould remarked that during his visit to Australia in 1838–40 he had observed two mature male lyrebirds 'chasing one another around, apparently in play'. Many visitors to Sherbrooke Forest will have observed this behaviour. The game usually begins when a bird from an adjoining territory ventures stealthily into his neighbour's domain. The latter becomes aware of the approaching bird long before either can see the other, and both birds may be seen moving their heads from side to side and up and down, as if trying to pin-point the exact location of the other bird. At this stage, both birds are silent (to human ears), but it seems certain that they are able to hear one another, possibly through an exchange of high-frequency notes. Suddenly, one bird runs towards the other and a chase begins; the birds course around, up and down the slope, for 10–15 minutes. The birds make a peculiar grunting noise like 'ugh…ugh … ugh…' and sometimes they make a chortling sound like 'eeaw…eeaw…eeaw…'; and the thudding of their feet on the ground may clearly be heard.

Sometimes a third bird takes part in what is obviously a game, because there is never any sign of hostility between the birds. At the conclusion of the game, the birds separate and often run to their mounds where they display, as if inspired by this brief contact with their neighbours. I have seen advanced immature males engaged in this activity, but it is more common with mature birds.

While the movement of the birds' heads begins when they are still unable to see one another, because of the dense vegetation between them, I have also seen lyrebirds engaged in such antics, with their tails pressed close to the ground and their heads moving so rapidly that the swish of air could be heard, even when they were close enough to be able to see one another distinctly. It seems that the 'snaking of the head' is a kind of ritual which is triggered by an auditory signal which the bird interprets as an invitation from a neighbour to 'come out and play'.

There is another peculiar form of behaviour exhibited by lyrebirds. A bird may be feeding in company with others, or it may be alone, obviously in harmony with its environment. Suddenly it utters a loud squawk and jumps up in the air, and then begins to course madly in a circle, bounding off the sides of five or six trees in each round, from heights of 1.5 or two metres, whilst continuing to squawk exuberantly. After a few minutes, it returns to its scratch patch and resumes feeding. The bird appears to enjoy this experience, which reminds one of a puppy or kitten at play, or a lamb frisking in the field. If the bird were alarmed, it would depart in haste and its note would be quite different. I can only conclude that this is a game which it plays on its own, as distinct from the game of 'chasey' played with others.

# Hearing and vision

The lyrebird is renowned for its acute hearing, which greatly frustrated pioneers like John Gould and A.J. Campbell when they were stalking lyrebirds in the wild. Too often, when striving to capture the song of the lyrebird with my tape-recorder, I have taken one step too many, causing the lyrebird (which could not possibly see me) to dissolve in the silence of the forest.

There is much evidence to suggest that lyrebirds are able to intercept signals outside the human auditory range. I have often seen a lyrebird engrossed in preening or scratching for food when out of the blue he has received a message from a distant source, although I have heard nothing. My hearing is very acute. The bird springs to the alert so quickly that the feathers he was preening have not the time to settle into place before he is straining in the direction from which the signal came, apparently trying to pin-point it. Sometimes, after a few moments of straining to locate the source of the signal, the bird will move off in that direction; at other times he may make a peculiar whimpering sound and slowly resume his former activity.

I have observed male lyrebirds on a mound 'straining' in the direction of the female sitting in the nest, 300–400 m away; obviously the birds were communicating with one another, but neither could see the other, nor was there any audible sound.

The large bulbous eyes of the lyrebird have no doubt evolved to ensure maximum efficiency in the winning of food in situations of low visibility. My experience with 'tame' lyrebirds like Spotty and Timothy of Sherbrooke Forest indicates that the lyrebird can see quite well at short distances, but evidence on the range of the lyrebird's vision is not so clear. I recently had an experience with three 'wild' birds which throws light on this matter. I emerged from a bend in a forest track in a remote area to see three mature male lyrebirds engaged in a game of 'chasey'. I was standing silent and stationary as the birds continued to run straight towards me. It was

*Dappled sunlight barely penetrates the dense vegetation of Sherbrooke Gully; in this secluded spot, the lyrebird builds her nest.*

*A mature male superb lyrebird in tail moult, in mutual display with an immature lyrebird.*

*Spotty in his favourite bathing pool, in Odell's Gully.*

*A lyrebird squeezing water out of his tail feathers with his beak, after bathing.*

*Spotty removing water from feathers under his chin, with his claws.*

*A female lyrebird preening after her bath in a creek at the end of the day.*

*Here the lyrebird is seen using the weight of his body to flatten an obstinate bracken stem near the edge of his mound.*

obvious that they had not seen me; but, when the leader was about 20–25 m from me, he suddenly stopped, uttered a loud alarm call and sprang off the road into the dense fern cover of the slope below. The others followed and I could hear them as they continued their game. I think this example sets the upper limit of the lyrebird's vision at about 25 m.

The hearing and vision of the lyrebird have no doubt evolved to meet the challenges of an environment of limited visibility. Acute hearing over a wide range of frequencies is necessary to enable the bird to communicate with other birds and to protect itself against predators, while the eyes need to be efficient at short range but the bird gains no advantage from having long-range vision.

*Spotty reacted instantly to a message received from a distant source.*

# Do lyrebirds talk?

The lyrebird is renowned for its mimicry of the birds of the bush. It is also claimed by various authors that the lyrebird's mimicry includes the sounds made by many animals such as the koala, the fox and the dog, and that its repertoire includes many mechanical sounds such as the blow of the woodsman's axe on the wood and the sharpening of the feller's saw, the rattling of chains, the hoot and 'choo choo' of the steam train, the surveyor's whistle, etc. Not every lyrebird covers the full range, the sounds imitated depending on local conditions.

It is not surprising therefore that the question is often asked: 'Does the lyrebird imitate the human voice?' Information on this aspect is limited. Ambrose Pratt (1932) stated that 'James', when greeted by his friend Mrs Wilkinson with a 'Hullo boy', responded in like manner. 'Jack' was a semi-domesticated free-ranging lyrebird which was taken from his nest as a fledgling, in 1885, and reared on a farm near Drouin in Gippsland. He is reported as having mimicked a variety of farm noises and to have encouraged one of the horses to greater efforts by calling out 'Gee-up, Bess', and to have repeated the warning 'Look out, Jack'. He is reputed also to have entered into debate with his farmer friends and, when referred to as 'Poor Jack', to have replied, 'Not "Poor" Jack, "Fat" Jack' (Godfrey, 1905).

I was disappointed that my friend Spotty never once attempted to repeat my greeting 'Hello, Spotty' and, on the whole, I accepted the reports concerning James and Jack with some scepticism. However, during a recent visit to the Healesville Fauna Park, I called on 'Chook', then in his sixth year. This bird was taken from his nest at the age of four to five weeks, hand-reared by Mr and Mrs Graeme George, and duly released into the lyrebird aviary. When I approached him, he indulged in a wing-raising display and made noises resembling a human voice, although no words could be identified. It may be significant that both Jack and Chook were imprinted to humans as fledglings; naturally, Chook will remain under close observation.

# Polygamy

It was long thought that the lyrebird was monogamous and that two lyrebirds having formed a pair-bond remained partnered for life (Pratt, 1933). The difficulty of obtaining reliable evidence on this aspect, under field conditions, can be imagined.

When in June 1960 I saw two well-known 'ladies of the forest' — 'Droopy' and 'The Landslide Female' — simultaneously visit Spotty on a mound, I was naturally curious to know whether he actually mated with either or both of them, but I did not see this occur. I have referred elsewhere (Smith, 1968) and on page 37 of this book to Spotty's interest in Droopy's 1960 nest, and it is common knowledge that Spotty mated with the Landslide Female for several years and that their progeny include 'Red' (hatched in 1958) and 'Blue/White' (hatched in 1963), both of which became well-known Sherbrooke identities in their time. In 1964 I saw 'Crossed Lyrates', at different times, courting two different females, about 100 m on either side of the Broadcast Area; however, in neither case did I observe copulation.

More conclusive evidence was obtained through the painstaking work of members of the Sherbrooke Survey Group who, by making observations on banded birds, were able to show that one particular male lyrebird courted and mated with three different females which laid eggs, during the breeding season of 1971 (Kenyon, 1972). Whether this polygamous behaviour is general among lyrebirds is not known, and it may be related to particular circumstances which resulted in the male's having domain over a large territory which embraced the territories of several females. In other areas, where the ratio of males to females is nearer unity, such behaviour might not occur. However, the fact that the male remains active for ten to twelve weeks would probably provide opportunities for polygamous unions.

# Predation

Throughout this book I will refer to the losses in lyrebird population due to predation. The red fox (*Vulpes vulpes*) is generally considered (correctly, I think) to be the principal offender, but reliable information on predation losses is difficult to obtain. Ward (1939, 1940) expressed concern at the growing threat to the lyrebirds in New South Wales. Birds of all ages are attacked, but chicks in nests are especially vulnerable.

Reilly (1970) found that, of forty-four eggs laid in Sherbrooke Forest during the period 1958–65, thirty-eight hatched, twenty-eight of the chicks later being identified through their colour bands outside the nest. This represents 63 per cent of the original eggs, and the overall result suggested that predation was not a serious factor.

On the other hand, Lill (1980) found that, in the south-eastern part of Sherbrooke Forest Park and in the Maroondah catchment area, near Healesville, of fifty-six eggs laid, the success rate was between 11 and 20 per cent. Predation accounted for 69 per cent of the losses. Among the possible predators, Lill listed foxes, domestic dogs and feral cats; losses from tree nests were thought to be due to kookaburras, currawongs or ravens.

The disparity between the success rates reported by these two authors seems remarkable. I have had no experience in the Maroondah catchment area, but am familiar with the south-eastern part of Sherbrooke Forest Park (this is where Tom Tregellas had his log home) and the area covered by the Sherbrooke Survey Group. Both areas are known fox resorts and it may be relevant that, during the period 1958–65, a vigorous poisoning campaign was conducted against the fox in the area covered by the Group. Numerous foxes were destroyed in such areas as Odell's Gully, Sherbrooke Gully, above and below the firebreak and Clematis Avenue walking track, etc. Unfortunately, some foxes remained at large.

The usual procedure was to obtain small chickens (casualties from a poultry farm or research laboratory), lace them generously with strychnine or 1080, and suspend them by strong cord from a low branch, within convenient reach of a fox. The fox therefore had to eat the bait on the spot rather than carry it away and perhaps drop it for some other animal or bird to collect later. We also used rabbits and rabbits' heads, with great success.

There may be some other explanation; but I am inclined to think that the results of Mrs Reilly and Dr Lill reflect the success of the fox-destruction campaign.

Brunner, Lloyd and Coman made a study (1975) of over 1800 fox scats collected from Sherbrooke Forest during 1973–74 and concluded, *inter alia,* that 'the absence of feathers recognizable as those of the lyrebird suggests that this species is seldom, if ever, preyed upon' (by the fox). Sadly, this statement is totally at variance with the facts.

# 3

# The male superb lyrebird

The body of the mature male superb lyrebird is similar in size to that of a domestic hen. The colouring varies somewhat throughout the range, being generally darker in southern parts than in northern districts. The head, hind neck, back and upper wing coverts are a dusky-brown colour, somewhat paler on the rump and tinged with grey. The breast and abdomen are a mid-grey colour (sometimes lighter, sometimes darker), while the wings are dark-brownish black, merging into rufous-brown towards their ends. The legs (tarsus) are very dark, almost black, about 114 mm long; the claws span 140–150 mm; the culmen is about 35 mm; weight is 900–1000 g, and the total length is about 1070 mm.

Campbell (1900) refers to an albino male lyrebird seen in Gippsland while Bill (1933) saw an albino female near the Baw Baw Plateau in Gippsland, and Kevin Mason of Badger Creek informed me several years ago that he saw an ashen-grey lyrebird at Mount Toole-be-wong. Lonsdale (1987) saw a *black*

The head of a superb lyrebird showing the large oval-shaped eye and nictitating membrane, which resembles a third eye-lid and is used for cleaning and protecting the eye.

lyrebird in the Hawkesbury sandstone country near Kurmond, New South Wales. However, these exceptions are rare.

# The display mound

The lyrebird is now recognized as an 'arena bird' (Rowley, 1974), a group of birds which includes the birds of paradise, bower-birds, ruffs and mannikins. Each of these species has its own courtship ritual.

The principal features of the lyrebird's courtship behaviour are based on the display mound and on his song. Here we consider the mound; the song will be discussed later.

To enhance his prospects of securing a mate, the male lyrebird constructs a number of

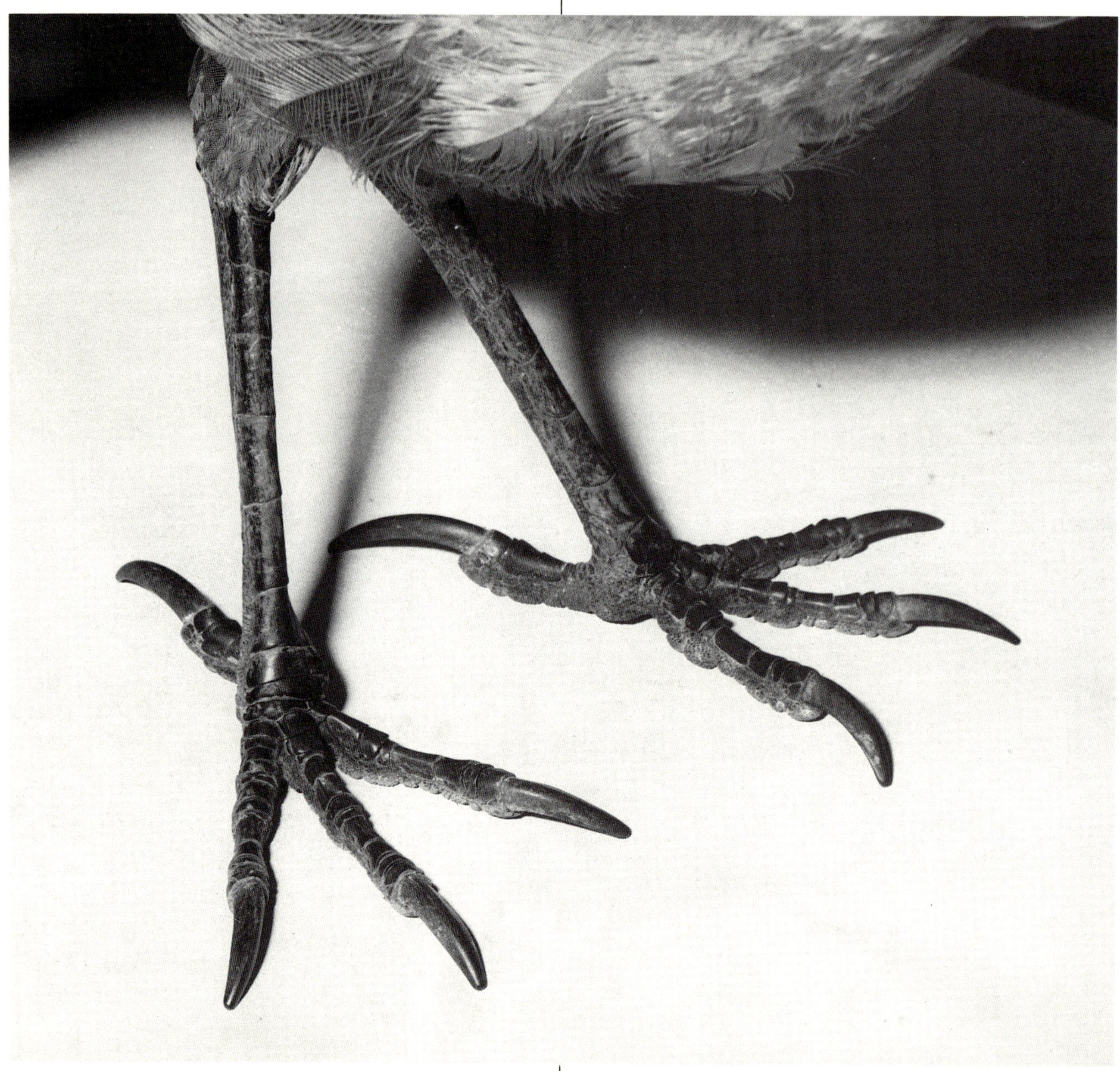

*The legs and claws of a mature male superb lyrebird; the claws span 14–15 cm.*

display mounds throughout his territory, which varies in size according to the nature of the district and other factors, such as lyrebird population density, and may range from about 1.5 ha to 3 ha. Just as the female constructs her nest unaided by the male, so too the male lyrebird builds his own mounds by scratching away the vegetation from the selected area with his long powerful claws. I have measured active mounds in Sherbrooke Forest, Tidbinbilla, Mount Buffalo, Gippsland and Tasmania, and regularly found them to be 1.7–1.8 m across, the middle of the mound being 10–15 cm higher than the perimeter. Presumably this slight elevation enhances the effect of the display; it would also assist in preventing the mound from becoming water-logged during heavy rains. The mound, of course, is at ground level — not, as described by one author, at heights of up to nine metres — and, although usually secluded, surrounded by ferns or bracken, it is almost invariably open to the sky above. This ensures that, when displayed, the tail will produce its maximum visual impact on the female lyrebird.

During the autumn and early winter, the lyrebird begins to refurbish his mounds of last year and sometimes makes new ones. In places like the ridges of Mount Buffalo and certain parts of Gippsland (low-rainfall areas) the ground is very hard and stony and the bird spends much time in raking the stones towards the perimeter so as to provide a central area of better tilth. This is important because in the course of the display he often pauses and scratches the earth towards the middle of the mound. In the spaghnum moss beds in Mount Field National Park, Tasmania, the tangled mass of roots makes it very difficult to construct a mound but, so great is the bird's need for it, he works until his goal has been achieved.

An individual bird may have as many as twenty mounds which he uses from time to time, but during the breeding season he may show a preference for a dozen or so of these, which he maintains in good condition. Some mounds are used for many years, even after the original owner has died or left the area. One particular mound in Sherbrooke Forest, in the Broadcast Area, was in active use for nearly forty years, but is now completely overgrown because there has not been an active male in that territory for several years. In the 1930s this mound was used by 'Silvertail', the magnificent male lyrebird featured in Ina Watson's book, *Silvertail — The Story of the Lyrebird*.



# Courtship behaviour

The mature male lyrebird may sing and display at any time of the year. This appears to be necessary for his courtship apparatus to be in peak condition at the appropriate time, when he will be required to compete with other lyrebirds. Even when he is undergoing a tail moult and before the new tail feathers have appeared he may sometimes be seen displaying on a mound, in the open forest, on top of a stump or on a log, to the accompaniment of his own song, but the performance lacks the fervour of a winter display.

During the non-breeding period song and display are not usually directed at any particular female. Especially at this time, displays may be directed at a stick, a fern or even a thick vine which happens to be hanging near the mound, but he may also be stimulated in various ways to engage in almost full winter courtship behaviour. A few examples will suffice.

I once watched Spotty (probably Sherbrooke's best known lyrebird) courting a plaintail, probably a female, on a warm blue-sky New Year's Day (1958), engage in 'whisper song', 'chortling', 'scissors grinding', 'clonk clonk clickety clickety click' calls and other vocalizations, whilst posturing at the plaintail with wing-raising, full-face and invitation displays.

There were several other male lyrebirds nearby and, from time to time, I could hear one of them approaching stealthily (the dry leaves on the forest floor were helping *me* this time!) but, as soon as the other bird reached a certain point, Spotty would rush at him and chase him off down the hill, and then return to the object of his passions, only to have to repeat the performance a short time later. On that day, Spotty showered his attentions on the plaintail for eight hours, without respite, but the female (?) appeared not to respond, although it has to be said that she did not attempt to escape from him, either. What makes this incident even more interesting is that Spotty's mate of the previous season built her nest only a short distance from the scene

*Spotty courted this female incessantly throughout a long summer's day (1 January 1958).*

of these activities, but I did not see either the female or the 1957 chick which later became better known as 'Spotty Junior'. A week later, when I found Spotty, he appeared to have no interest in the opposite sex.

Although the male lyrebird is sexually quiescent during the spring, summer and early autumn, nevertheless he may, if suitably stimulated, endeavour to mount a female or plaintail. On 11 November 1967 I was watching an immature male lyrebird as he sang and displayed on a mound. He had not yet acquired his first filamentary feather and therefore resembled a female. Suddenly, I saw Crossed Lyrates (a mature male which had earlier in the season been active in the Broadcast Area) quietly emerge from the ferns

*Spotty in a 'wing-raising' display to a female lyrebird (1 January 1958).*

and move towards the mound. Crossed Lyrates was in tail moult and therefore sexually inactive, but the sight of that bird on the mound in a position similar to that adopted by a female inviting coition raised the hopes of the gay troubadour, causing him to attempt to mount the plaintail. The latter, of course, was not responsive and indicated this to the amorous one; but Crossed Lyrates, undeterred by his rejection, persisted long enough for me to take two good colour shots of the incident.

I have seen immature male lyrebirds attempting to mount plaintails, and the mutual display of females (page 106) is another example of homosexual behaviour exhibited by many other birds and animals (Terres, 1980).

The untimely onset of cold foggy conditions, even in February (months before the breeding season), may stimulate a mature male lyrebird to vigorous courtship behaviour. On such a day, when the forest was enveloped in fog and light drizzle, I watched a mature male from the Wattle Walk area (over a kilometre away from the scene of the incident being described) courting a female which was accompanied by a chick from the previous season. He treated her to the whole range of vocal and display activities in the courtship ritual, but the female merely kept moving away from him as she tended her chick. It was quite apparent that she was not 'at home' to him, but he persisted for several hours and was still pressing his suit when the feel of cold water trickling down the back of my neck caused me to move on.

As autumn gives way to winter, the tempo of courtship activities continues to rise. The male now courts the female assiduously, following her as she feeds, seemingly oblivious to his advances but no doubt stimulated internally. The male follows the female with his tail feathers trailing and neck outstretched, continuously engaging in whisper song, chortling, scissors-grinding and mimicry (especially the subdued chuckling of the kookaburra). Throughout this vocal courtship he indulges in wing-raising and frequently stands on one leg, with the other delicately poised in characteristic fashion, while he engages in whisper song with his beak only slightly open. Usually after 15–20 minutes of this behaviour he will run to a log or stump on which he displays. When he begins to make a continuous 'blick blick blick blick...' note, while holding his tail in an almost erect position, one knows that he is about to run to a mound, sometimes 100–200 m away. Here he will display for perhaps 30 minutes or he may leave that mound after a few minutes and run to another... and another, before settling down to a prolonged display, which may last for 45 minutes. After this, he may take a rest from displaying and resume feeding. It seems safe to conclude that all the activities of the past hour or so have depleted his energy which must now be replenished. Prolonged singing

also leaves him thirsty and usually he will visit one of his favourite drinking spots. A depression in a strip of bark or in a large log, or the saucer-shaped top of a large fungus in which the overnight dew or rain has collected, will suffice and sometimes saves him a long walk.

Not every display is preceded by the courting of a female at close quarters and after the birds have mated the female is seldom seen near the male, except perhaps by accident when their foraging excursions occasion a chance meeting. At such times there is instant recognition; the male vibrates his tail feathers and may even sing quietly and display at the female, while she acknowledges him by moving her tail from side to side.

This pattern of behaviour continues, the two birds no doubt being stimulated by the mutual interaction, until they are ready for the final act of mating. This usually occurs early in June, extending into July and, occasionally, into August.

In the course of the display the sixteen tail feathers are inverted and brought forward over the bird's head so that the body is often concealed. When the tail is displayed the viewer is presented with the spectacle of dazzling filamentary plumes which may be fully extended (as in a full-face display) or raised like a fan. If the female is near the mound, the male may direct a special effect at her by vibrating the feathers rapidly, creating a shimmering effect. When the tail of the superb lyrebird is fully displayed it covers an area of about 1.5 square metres. Of course, whether the female interprets these activities in the same manner as a human observer is open to question, but it is certain that the exercise is performed for her benefit, not for the pleasure of any human being who, through sheer good fortune, might be an interested spectator.

Frequently, in the course of the display the bird performs what has come to be called 'the dance'. The bird moves his body rhythmically from side to side; he may move backwards, sideways, then forwards and back to his original position. Then the performance may be repeated, perhaps several times. This movement is usually accompanied by the 'scissors-grinding' note. This is difficult to describe but it sounds like the scissors-grinding notes of the restless flycatcher. As a variant, the bird may jump from one foot to the other, to the accompaniment of the 'clonk clonk clickety clickety click' call.

In another feature of the display, the tail feathers are held aloft and slightly forward, and the wings are drooped.

The 'invitation display', which is an important part of the mating ritual, has already been referred to. Displays often commence on top of a stump or log, or in a tree where the bird has been singing. He may even attempt to perform the dance on the limb of a blackwood tree or mountain ash, 16 m above the ground, but soon finds himself cramped for space and rapidly descends from his arboreal stage to display on a mound.

Sometimes the female plays 'hard to get', possibly because the male has moved closer to his peak of sexual activity than she has and she is not yet ready to accept him. At all events, she frequently leads him on a chase, through the ferns, up into the trees, from branch to branch and, as the male seems set to catch up with her, she glides gracefully to earth, followed after a brief spell by the baffled male, and the chase is resumed. Usually the chase ends when the male, thoroughly stimulated, runs to a mound and displays.

At the end of a display on a mound, in the absence of the female, the bird usually pauses for a few minutes, with tail half erect or folded behind him and beak almost closed, imitating the scrub wren. When he ceases, he raises his crest and 'bows out' after the manner of a great artist. Sometimes he returns as if to respond to the plaudits of an invisible audience.

# Lyrebirds fighting

When a lyrebird is quietly scratching on the forest floor, seeking food, he is often attended by one or two robins, a pilot bird or a few scrub wrens. These little birds sometimes alight on the larger bird's back or hop between his legs as they steal his food, and are sometimes accidentally trodden on. Once, I saw a robin caught in the great claw of a lyrebird as he reached out to scratch. Happily, the little bird was unhurt and was soon back in the worm patch. I have never seen a lyrebird show any sign of aggression towards the small visitors to his meal table. But if any of them feels tempted to hop on to his mound while he is displaying, the lyrebird shows his displeasure by moving his body forward in an intimidatory gesture which causes the small bird to depart in haste. To the lyrebird, his mound is holy ground!

However, at times, lyrebirds are much less charitable towards their own kind. Sometimes, even during the non-breeding period, one lyrebird will 'take a set' against another, possibly because of their mutual, if temporary, interest in a female, but the reason is not clear. I once saw Spotty drive a young male lyrebird into the protection of the dense accumulation of dead fronds at the base of a large tree-fern. Whenever the younger bird endeavoured to break away Spotty menaced him and held him captive. The war of nerves continued for a quarter of an hour until Spotty, possibly because of hunger, momentarily lost concentration, thereby enabling the prisoner to escape. On another occasion I was watching Timothy and a young male lyrebird as they competed for the favours of a plaintail (probably a female). Suddenly, Timothy gave the young male a solid push with his claws, sending him sprawling. One afternoon in September 1947 Timothy was displaying on a mound while another male, which often joined Timothy in a game of 'chasey', stood quietly near the edge of the mound watching. There was no hostility here; these two birds were old friends. But, suddenly, a third lyrebird, in complete tail moult, crept to the edge of the bracken and sprang on to Timothy's mound. The effect was electrifying. Timothy gave a loud angry squawk and set after the intruder. They coursed around the forest for several minutes during which time I could hear the thudding of their feet. I am certain

*Spotty pauses in his quest for food to release a burst of melody, while two yellow robins keep a hopeful eye on his worm patch.*

that if Timothy could have caught the bobtail lyrebird he would have punished him severely. It seemed strange that the offender made no attempt to fly, which in the absence of his tail he could easily have done. Perhaps in his fear, and because he so rarely used it, he forgot that he possessed this skill. He was obviously in distress and was beginning to wobble as he ran. Fortunately, at this moment Timothy abandoned the chase and ran about 100 m to one of his favourite mounds, where he sang and displayed for half an hour, without further interruption.

In the early 1950s the Broadcast Area was occupied by a mature male known as 'The Wanderer', which was readily identifiable by virtue of a white pigmentation fault on the left flank. One Sunday evening, just before dusk, he was quietly preening on a low perch before

This yellow robin was accidentally caught in Spotty's large claw, but escaped unharmed.

going to roost when, suddenly, he ceased preening and peered intently into the darkening forest ahead of him. Then he uttered a note of annoyance, glided to earth and began a murderous pursuit of a younger bird which had innocently intruded into his domain. In desperation the younger bird bolted down a wombat's burrow and his pursuer stood guard at the entrance, strutting to and fro, swishing his tail angrily and making loud harsh guttural noises, the like of which I had not heard before, nor have heard since. After a while, perhaps encouraged by the owner of the burrow, the younger bird shot out of the burrow like a rocket and glided down into the dense fern gully, with the baffled Wanderer in hot pursuit.

Sometimes, however, lyrebirds do actually come to blows, and clusters of feathers occasionally found in the forest are evidence that one bird has raked another's breast with his claws. R.C. Chandler, writing over eighty years ago of the lyrebirds in the Bass Valley, related having seen two male lyrebirds fighting like roosters, using their beaks and sharp claws and sometimes tripping over their tails.

Several years ago, Richard Brown, of Nunawading, Victoria, informed me that on 31 May 1965 he had observed a 'changeling' lyrebird make a sudden attack on a 'plaintail'. The latter screamed and struggled, and was thrown on to its back, in which position it was held with one claw by the other bird while he raked feathers out of the victim's breast with the other. The aggressor tried to peck the other bird and at one stage seized its beak in his. The two birds remained in this position for some time, the plaintail screaming and struggling occasionally, while the other one was silent but continued slowly to rake the other's breast. There were several other lyrebirds nearby, but they showed no concern, except when the plaintail screamed. Eventually, the changeling released his hold and the other bird escaped, only to be chased over logs and through the ferns for a further five minutes, after which both birds resumed their feeding.

To this point, my knowledge of lyrebirds fighting had been hearsay. However, my experience was soon to be broadened; because on 19 May 1967 I was watching a mature male lyrebird displaying on a mound in Sherbrooke Forest. It was about 10 a.m., and he had already displayed on several other mounds during the morning. Suddenly, he swung around and faced the east; the rapid vibrations of his tail suggested that he had become aware of a female nearby; and then I saw a movement in the ferns some distance away. Soon, a plaintail came into view and then, at the clear 'invitation' of the displaying male, walked on to the mound. In a flash, the male bird sprang at the new arrival and knocked him on to his back. For, indeed, it was a young male that the gay courtier had invited on to the mound. The young bird, probably in his fourth year, squawked and released several ear-piercing screeches in quick succession, but then became silent as he strove to avoid the deadly thrusts of the attacker's beak, aimed like a dagger at the victim's head. The young bird, still on his back, endeavoured to pull away and somehow managed to wrap his claws around the thighs of the older bird, thereby preventing the latter from reaching him with his beak. So the two birds lay there, locked together by their legs and claws, for several minutes, in complete silence. Eventually, the strain proved too much for the older bird, which released his grip. In a flash, the other bird did the same and raced, somewhat erratically, from the mound, hotly pursued by the older bird. I could hear them threshing about in the ferns, but the youngster was free for the time being.

As I stood watching the birds in combat, my thoughts flew to 'One Eye', an immature male lyrebird which used to frequent the same area, and I felt then that I knew how he had lost his eye. Poor One Eye ... eventually he was struck and killed by a passing car, as he was crossing Sherbrooke Road, a few hundred metres from the scene of the combat I had just witnessed.

The nearest I ever came to seeing a fight between two mature male lyrebirds occurred about a year after the incident just described, when two male lyrebirds (one of whom was the aggressor in the combat just referred to, and the other was one of Spotty's progeny) came into conflict over a small piece of worm-rich earth. As Spotty's offspring made a move to take over the worm patch from the other bird, the two birds sprang up into the air, almost standing on their tails, and struck one another with their claws. When they came to earth, Spotty's offspring decided that discretion was the better part of valour and moved away.

# Nesting behaviour

Although Le Souef (1903) claimed to have seen a male lyrebird sitting on the egg in a nest near Gembrook, Victoria, no confirmation has been forthcoming. It is now accepted that, in the wild, the male lyrebird does not participate in the building of the nest, in the incubation of the egg or in caring for the chick. Various reasons have been proposed for this behaviour, the principal ones being that the display and vocal activities of the male would attract predators to the nest. I think there may be a more fundamental reason.

Until the advent of white settlers in Australia, the predators with which lyrebirds had to cope were native cats, dingos, possibly possums, hawks and owls. If the predators of those times had any appreciable effect on the lyrebird, at all events, the latter managed to survive and develop in such numbers that they could subsequently be shot by the thousand, and nests could be found with comparative ease. It seems to me that the white man's introduction of the European fox (*Vulpes vulpes*) and the domestic cat (which has been permitted to proliferate as a feral animal) has had a far more serious effect on lyrebirds than any of the other species mentioned, but the nesting behaviour of the lyrebird must surely have been established long before the fox and cat arrived.

I think that the real reason why the male lyrebird does not participate in nesting activities is that the length and structure of his tail preclude it. If the male lyrebird, at some time in the historic past, ever did take part in nesting activities, the break came when and if the structure of the nest changed (possibly in response to changing climatic conditions), because it is obvious that, even if the male lyrebird succeeded in getting into the nest (as we know it today and as it has probably been for a long time), the glittering tail filaments would project far beyond the entrance, thus destroying all semblance of camouflage. Moreover, it is equally certain that any attempt by the male lyrebird to enter the nest, by virtue of the comparatively rigid structure of the basal part of the tail feathers, would result in the destruction of both nest and tail. The participation of the male in nesting activities would demand a much larger nest, with all its attendant disadvantages such as loss of se-

clusion, time of building, heat losses, etc. So, if the two sexes ever did co-operate in nesting activities at some time in the distant past, they reached a point at which the female was left to perfect the construction of a nest suited to the needs of the species, while the male was left to develop his tail to enhance his biological efficiency.

There has been some debate as to whether the male lyrebird is aware of the location of the nest and whether he will defend it against predators. Droopy once built a nest in Spotty's territory and presumably they mated. Spotty was observed placing his head in the nest, which suggests that he had a special interest in it. However, the issue is clouded by the fact that Spotty had a predilection for holes. When passing a certain wombat's burrow on the eastern slope of Odell's Gully (which he often did), he was in the habit of entering the burrow and remaining there for 10–15 seconds, sometimes longer. Nor was he the only lyrebird seen to do this.

Once, in Odell's Gully, I heard a mature male lyrebird imitating the noises made by his own chick while it was being fed by its mother; this suggests that the male bird was aware of the location of the nest. Reilly (1970) showed that some male lyrebirds are well aware of the location of their respective nests and that they engage in distraction behaviour if the nest is approached by a human being. However, Lill (1979) reported that 232 female alarm calls resulting from his inspections of their nests evoked only 1.3 per cent of male lyrebird responses. Lill suggested that the males responded because they just happened to be near the nest in the course of their foraging at the time of his visits to the nests.

Whether the male lyrebird would, in any circumstances, defend the nest against predators seems very doubtful. When I have approached their nests, I have often been vigorously dive-bombed by angry females, but never have I seen a male lyrebird defend a nest.

There are two cases on record which show that under certain conditions the male lyrebird may be induced to care for the young. Fleay (1941) reported that early in January

1941 the driver of a timber truck found a young lyrebird at the edge of a forest road and took it to the Sir Colin MacKenzie Sanctuary at Healesville, Victoria, of which Fleay was then Director. Fleay placed the two-to-three-month-old chick in an aviary containing a mature male and a mature female lyrebird both of which had been saved from bushfires in the district. Neither bird was at that time in breeding condition. Within two days the soft cheepings of the chick had induced the female to adopt it and to search throughout the entire day for worms, grubs, etc. and store them in her cheek pouches before transporting and feeding them to the chick. Two weeks after the arrival of the chick, the male also had adopted it and outclassed the female by his attentiveness to it. His cheek pouches became distended just as do those of the female, to enable him to store food as he was foraging, and his enthusiasm for this unaccustomed duty caused him to feed the female lyrebird as well as the chick.

In 1985, the mature male and female lyrebirds in the Healesville Sanctuary mated and a chick was duly produced. Unfortunately, three months after the chick hatched, the female lyrebird died. As the chick was not yet self-supporting, the head keeper, Kevin Mason, assumed the role of foster parent and undertook the tedious task of feeding the chick. However, he was greatly relieved when, after nine days, the male lyrebird responded to the chick's persistent begging and cheepings. The male lyrebird gathered food and stored it in his cheek pouches, just as the female had done, before feeding it to the chick, which he successfully reared. Happily, the young bird is still alive and well at the time of writing.

There are interesting differences between the two cases described. In the first case, there was no parental bond between the chick and the adult birds, while in the second case the adult male was the chick's parent; yet, after some delay, both males had adopted the respective chicks. There is little doubt that the strange behaviour of the male lyrebirds was induced through the activation of their endocrine systems in response to visual and aural stimuli from the chicks, and it is probably significant that the female responded to the chick's signals within two days while the males required more prolonged stimulation. As mentioned elsewhere, under normal conditions the feeding of the chick by the female is a natural step in an hormonally directed chain of events; but the evidence suggests that the female in question, having lost her natural home in a bushfire, was not already programmed to embark on an adoption process, and certainly not in January. Yet she responded within two days.

However, the major difference between the two cases was that, in the second example, the male lyrebird carried the full responsibility of successfully rearing the orphaned chick, without the stimulus provided by the sight of his mate regularly responding to the chick's persistent demands.

The feeding techniques adopted by the male lyrebirds (even to the extent of storing food in the cheek pouches and carrying it to the respective chicks) were identical with those of the female and it seems that the hormonal basis of the feeding behaviour is similar for both sexes, but the male birds required a much longer period of stimulation than the female did. I am not aware of any evidence which would suggest that under natural conditions the male lyrebird would undertake parental duties in the event of his mate's death, but it seems that the circumstance of being confined within an aviary imposes additional pressure on the male, causing him ultimately to respond to the unusual external stimuli. This leads to the broader question of whether there are latent within both male and female lyrebirds vestiges of behavioural characteristics of the opposite sex which may, under conditions of stress, be activated through the endocrinal chain.

*A side view of the 'invitation' display.*

*A frontal view of the 'invitation' display.*

*A mature male lyrebird in tail moult attempting to mount an immature male.*

*A mature male superb lyrebird courting a female, early in the breeding season.*

*A mature male lyrebird engaged in pre-nuptial courtship and 'whisper song', early in the breeding season.*

*Spotty giving his territorial call, whilst perched on a favourite log.*

*A mature male lyrebird displaying, facing directly into the camera.*

*A rear view of the display, showing the symmetrical arrangement of the tail feathers.*

4

# The female superb lyrebird

The female superb lyrebird weighs about 900 g and is generally similar in size and colour to the male, but her tail is shorter and consists mainly of plain feathers — twelve main rectrices (plain feathers), two plain medians and two lyrates. The main rectrices range in length from about 300 mm at the outer side of the tail to about 400 mm at the middle; the medians are about 440 mm long and highly asymmetric, and the lyrates are about 305 mm long. All tail feathers are brown to blackish-brown on the upper side and silvery-grey on the underside. The main retrices and medians usually (but not invariably) have rounded ends, and the lyrates have orange-chestnut bars and 'windows', but the apices of the Vs do not extend to the shaft and are absent from the distal 50–70 mm of the feather, which also lacks the black tip of the male lyrate. In general, the female lyrebird resembles an immature male which has not yet acquired a specific male characteristic.

Information on the age at which a female lyrebird matures is limited. The only information I have pertains to a bird which was banded White/Red as a nestling in August

*An adult female lyrebird, showing the rounded ends of the twelve plain feathers and two medians and the large bulbous eyes. The windows in the lyrate feathers do not extend to the shaft.*

1963 and was observed incubating an egg in a nest in the eastern part of Sherbrooke Forest Park in August 1970. Thus, she matured in her seventh year. (See also page 104.)

The adult female lyrebird is one of the most industrious members of the avian world. She alone builds the nest, incubates the egg, cares for the chick in the nest and later in the forest until it becomes self-sufficient, at which stage she embarks on a whole new round of domestic chores, so that she is never free of her obligations to the species.

# Territorial behaviour

The adult female occupies a territory which, however, is smaller than that of the male and lies at least partly within it. My experience, which appears to be confirmed by that of Reilly (1970) and Kenyon (1972), suggests that there is some benefit, from the female's viewpoint, in having a territory. She needs access to nesting material and to food for herself and her chick, and it must be advantageous for her to be close to a mature male.

Certain Sherbrooke females have been known to spend most of their time, especially during the breeding season, in particular areas, to the almost complete exclusion of other females. The term of occupancy depends, I think, very much on the luck of the bird in avoiding predation by introduced animals. For several years (1947–60) one female regularly built her nest in an area just to the north of 'the Falls', while another built on the opposite slope. Another bird, Droopy, usually built on the eastern slope of Odell's Gully, while the Landslide Female invariably built near a small landslide which occurred about forty-five to fifty years ago.

The female defends her nest against intruders and some birds behave very aggressively. When I attempted to inspect my first lyrebird nest in 1938, the female launched herself from a nearby tree-fern and flew directly at me, trailing her claws through my hair. She squawked loudly and continued to attack me until I retreated, but I soon learned

that, if I remained motionless and did not approach the nest too closely, she would settle down and resume her quest for food, while still keeping a watchful eye on me. After several visits, I had won her confidence and, despite the poor light, managed to obtain a good portrait of the bird — my first. Several years later (1948), in the same area, I made the brief acquaintance of another female which attacked me unceasingly, using her nest as a launching platform for her aerial sorties, causing me to abandon all further interest in that nest, to save it from complete destruction.

Other females, however, have become so 'friendly' that I have been permitted to sit quietly by the nest and even have visitors there. Sometimes, when I was squatting near the nest, she would scratch right up to my boots and virtually request me to move aside, under penalty of having the soil scratched out from under my foot. The secret of winning a lyrebird's confidence is to avoid rapid movement and to allow the bird to dictate the terms. Tom Tregellas described similar experiences.

However, the female lyrebird can at times be very aggressive towards another female which ventures into her domain and occasionally threat displays between female lyrebirds lead to fighting. Female lyrebirds often remove nesting material from a nearby nest in course of construction, though more frequently from an old nest and, on rare occasions, actually destroy another bird's nest in an adjoining territory (Reilly, 1970).

Territorial rights have little significance during the non-breeding period; then females cross one anothers' territories and often wander far from home, accompanied by their last-season's chicks. I have often seen females from the Wattle Walk, Landslide, Sherbrooke Gully and other areas in the forest above and to the east of the firebreak area, along with their chicks. This area is also a popular rendezvous for immature birds and mature males during the non-breeding period; but during the breeding season it is usually the territory of one particular male, and several nests have been found there.

*A female lyrebird giving a threat display near her nest.*

# Courtship and mating

While the male engages in a vigorous campaign of song and display over a period of several weeks, the female remains passive and responds outwardly only by a modest and somewhat coy swaying of her body and tail. Internally, her endocrine system moves fairly rapidly to a peak of short duration, which prepares her to accept the male. As she normally lays only one egg each season, the timing of copulation is vital. Fertilization occurs by simple cloacal contact, climaxing a gymnastic performance of some merit, and the sperm passes into the oviduct and fertilizes the egg shortly after it has been released from the ovary. If fertilization does not occur, the egg

may still develop, but will not produce a chick.

As a prelude to mating the female goes under the tail of the displaying male.

# The nest: design and construction

The lyrebird's nest is highly functional in design and built to specifications dictated by climatic and other conditions. Because the breeding season occurs during the coldest period of the year, when temperatures in some parts of the Dandenong Ranges are often close to or below freezing point, and rainfall is usually high, the nest must be well insulated

*The nest of the superb lyrebird, showing the relevant features: A an outer nest of sticks; B the inner nest of fine fibrous material; C the platform or step, to facilitate entering and leaving the nest, and the performance of nesting duties by the female; D the spray of gum leaves for camouflage. This nest was built between the buttresses of a large mountain ash, about one metre above the ground.*

and waterproof. The large size of the female (about 700 mm long), having a tail about 450 mm long, and the need to provide accommodation for both female and chick for some four and a half to five weeks, demand that the nest be commodious. And the fact that it must withstand wear and tear for thirteen to fourteen weeks in addition to any period which elapses between the completion of the nest and the laying of the egg requires that it be durable and built on strong foundations.

The nest has been described by A.J. Campbell (1884) and Reilly (1970), but here we look more closely at the reasons underlying nest design and construction. The nest consists of three parts: a firm base (cradle), an outer nest of sticks and an inner nest of fine fibrous material. Regardless of the nest site, the female first scratches away any rubbish or, if she is building on a bank, she scratches out a niche in which to build a cradle or platform. The cradle is built of stout sticks up to 19–20 mm thick, all tightly interlaced. Next she builds up the sides of the nest, using thinner, more pliable sticks laced tightly together and tied into the cradle. To reduce heat losses and to assist in waterproofing the outer nest, the female plugs the interstices with moss and leaves. The inner nest is composed of fine fibrous material (tree-fern roots), dry bracken fronds, wire-grass, leaves and moss, all tightly packed.

In manipulating her building materials, she uses her beak, claws and body (especially her breast and shoulders) to form the material into the required shape. The female often collects the lining from an old nest, and some birds are not above accepting a helping hand. In June 1960 while watching Droopy building her nest I placed an old possum's nest in her path. Within two days she had incorporated it in her nest.

External dimensions vary, height and width ranging from 60 to 75 cm, the depth being about 50–70 cm. An entrance platform is formed by extending the base of the stick nest for a distance of 12–15 cm, the gaps between the sticks usually being plugged with moss. This greatly facilitates the female's entry

to and departure from the nest and is especially valuable in providing her with a secure platform on which to stand while she is feeding the chick and performing sanitary duties.

The upper part of the nest is enclosed in a 'hood', designed to reduce heat losses and protect the occupants against inclement weather. To reduce heat losses further, the entrance is at the side (or front) and is restricted to 15–20 cm, giving the sitting female a commanding view, while the nest is inconspicuous or invisible from the rear. In effect, the inner nest is a tunnel about 25–30 cm long (from front to back) with a dome and a slight depression towards the rear; this gives added capacity and ensures that the egg will not roll out. The cavity is about 25–30 cm wide and high.

The female always enters the nest head first and turns around inside. The interior having been designed so compactly, the female is obliged to fold her tail back over her head and to the side so that no part of her projects beyond the entrance while she is sitting in the nest. To do otherwise, as one naturalist put it, would be 'to frustrate the art of concealment'. Because of her long occupancy of the nest, the female's tail feathers become curved to one side during this period, which sometimes helps one to find a nest.

The domed structure of the inner nest enables the female or chick to stand up in the nest and hide itself, so that only two stick-like legs would be visible to any passer-by (human or animal) should the occasion arise. It also provides space for the growing chick to stand up, preen, stretch and exercise and relieve the monotony of sitting for six to seven weeks.

Finally, the keynote of nest architecture is camouflage. This can be so effective that one may be standing alongside a nest without immediately being aware of it. There is an amusing story told by H.C. Rawnsley (1863) who, towards the end of a long search for a female specimen, steadied himself against 'a heap of sticks' while he was taking aim. Imagine his surprise when the shrill alarm call of a lyrebird chick assailed his ears

*A female lyrebird with a bent tail, caused by sitting in the nest.*

from within his support. The camouflage is enhanced by the addition of a spray of gum leaves across the top of the nest. This appears to be a species characteristic, because I have observed it in many widely separated parts of the range.

Many lyrebirds begin the construction of a nest and carry it to the platform or cradle stage, only to abandon it and build elsewhere; or they may return to it a few weeks or months later, even two years later, and complete it. Howe (1927) describes a platform nest near Selby in 1917. When Tom Tregellas photographed it during the winter of that year, it was unchanged; but on 13 July 1918 the nest was complete and contained an egg. In another case (Reilly, 1970), a nest built to the platform stage was found, over five months later, to be complete and to contain an egg.

I have found several cradle nests in Sherbrooke Forest and in March 1981 found one at the base of a mountain ash at Mount Toole-be-wong, near Healesville. No further work was done on the platform that year; it was built up a little in 1982, but was completed in 1983. When I inspected it on 13 August, it had been feathered in preparation for the laying of the egg, but had been pre-dated. Feathers of the female found nearby suggest that she had been taken by a fox just before she was due to lay.

There appears to be no known reason for these platforms; they might be the work of immature birds.

While in the design of her nest the female lyrebird has shown herself to be very practical, in the selection of a nest site, she appears erratic. Nests have been found on or near the ground, at the base of trees or tree-ferns, on large logs against a tree, on stumps of various heights up to seven to eight metres, in the fork of a large musk tree or mountain ash, high up in a blackwood tree, in the crown of a tree-fern, on the bank of a creek and in various other situations.

Although a nest close to the ground offers advantages to the lyrebird, in so far as building efficiency and care of the chick are concerned, the higher nest is safer from foxes although still vulnerable to owls and cats, and perhaps possums. I have known a possum to eat young blackbirds in the nest and even take over the nest as a home. However, most nests are built reasonably close to the ground, as the following table shows (Reilly, 1970):

| Height of nest above ground (m) (N = 62) | Percentage of nests at this level |
|---|---|
| 0 – 1.2 | 60 |
| 1.2 – 2.4 | 21 |
| 2.4 – 3.6 | 6 |
| 3.6 – 18 | 13 |

High nests warrant special consideration from three particular aspects, though there may be others. Firstly, there is the aspect of nesting efficiency, by which I refer to the use made by the lyrebird of the time occupied in nesting duties; secondly, there is the aspect of safety from predators; and thirdly, there is the aspect of the safety of the fledgling during the descent — its first experience outside the nest.

The following 'time and motion' studies illustrate some of the disadvantages of the high nest, in so far as building efficiency is concerned. On 7 June 1978 I was in the eastern part of Sherbrooke Forest Park when a solitary 'aw-kok' call sent me hurrying down the slope into the dense forest. I soon located a female engaged in building a nest near the jagged top of a large mountain ash which had lost part of its crown in a storm several years ago. She had made considerable progress with the stick nest and was carrying the inner-nest material. After filling her beak with nesting material, she made her way to a particular tree about 13 m from the mountain ash and sprang on to a branch from which she began her zig-zag flight from branch to branch, until she reached the nest, 24 m above the ground (by clinometer). However, she did not always fol-

*A female lyrebird carrying nesting material.*

low the same route and sometimes found herself in an awkward situation, causing her to lose time while she extricated herself and found a way through. Having made the addition to the nest, she sometimes glided smoothly to earth but at other times the descent was prolonged as she worked her way through the labyrinth of branches. I made records of the time taken to reach the nest, the time spent at the nest (working time), and the time spent in descending. Over a period of 51 minutes, she made seven visits to the nest at intervals of 5, 14½, 5, 8, 8½ and 6½ minutes; the time to reach the nest varied between 85 and 140 seconds; the descent took from 6 to 30 seconds, and the times spent at the nest were 240, 90, 5, 5, 6, 10 and 5 seconds, a total of 361 seconds, or 12 per cent of the total time. She then ceased work; later, I found that the nest had been abandoned, though I do not know the reason.

This bird appeared to be lacking in nest-building experience and contrasted sharply with another which I had observed in Odell's Gully on 11 June 1955, building in the crown of a tree-fern, about four metres above the ground. This bird worked unceasingly, running all the time she was on the ground and invariably following the same route to the nest, to which she paid twenty-four visits in the course of one hour. Reilly (1970) described a bird which worked assiduously for two and a half hours, carrying nesting material to the nest on an average of once every six minutes; and on the following day she worked for two hours, visiting the nest every four minutes. This nest was about five metres above the ground.

In 1941 two of the nests in Sherbrooke were built at heights of 21 and 27 m respectively (Campbell and Gray, 1942). I did not see these nests being built, but the birds must have been under great stress, especially when carrying heavy sticks for the outer nest. Later, when the chicks had been hatched, I used to stand fascinated by the resolute manner in which the devoted females made their way upward from one branch to another in nearby trees until they were high enough to fly or spring across the intervening space to the nest to feed the chick and, in my memory's eye, I can still see those birds gliding gracefully to earth to resume their search for food. In June 1966 I saw a female fly to the top of the lower of these two dead trees, which had been used twenty-five years previously; but, although she made several visits to the site, she did not

proceed with a nest, and I must confess that I was a little disappointed.

The reason why some lyrebirds build their nests at a considerable height has been the subject of controversy and remains conjectural, but there are some who see it as a means of escaping from the introduced European fox. There is much evidence that the fox has had and continues to have a devastating effect on the lyrebird and on other ground-dwelling species.

In 1905 A.E. Kitson wrote, 'The days of the lyrebird are numbered unless it develops the habit of nesting in trees or other spots inaccessible to the introduced European fox', and in 1916 L.C. Cook reported that the lyrebirds in the Poowong district of Gippsland used to build on the ground, but 'since the fox came, they have begun to build in trees'. Gregory Mathews (1918), in his *Birds of Australia*, quotes from a letter he had received from E.J. Christian, who stated that 'for the past few years, many nests have been found up in trees, so they [the lyrebirds] are getting out of reach of dogs and foxes'.

But not every student of the lyrebird agrees that the fox is responsible for having caused the lyrebird to build her nest in high places. In 1919 A.V. Edwards reported having found lyrebirds' nests in all the situations mentioned above, including the crowns of tall trees and one in the fork of a tree 60 m above the ground, long before the incursion of the fox into the area in question, namely, the outskirts of Tantawanglo Mountain, 400 km south-west of Sydney. Edwards asserted that 'the lyrebird is naturally erratic in her choice of a nest site'.

Support for this view was provided by J.G. O'Donoghue (1931), who observed that, fifteen years previously, before the rabbit and fox had penetrated into the Crooked River district, 'lyrebirds were as common as sparrows and starlings in the suburbs of Melbourne', but that 'now you would not hear or see one in a week's ramble', while 'the valley teems with rabbits and foxes'. Yet he quotes A.W. Milligan (who had had much experience with the lyrebirds of that area and who had experimented

*A female lyrebird flying away from her nest in a tree high above the ground. She is carrying the chick's dropping in her beak.*

with hybrids between lyrebirds and the domestic fowl) as saying that 'the assumption that the female nests above the ground because of the fox is erroneous' and that 'the hen bird would have to sustain many losses in eggs and chicks before it became obvious to her that high-nesting was the only expedient by which she could rear her young'. This view was supported by George Mack (1952).

It would seem that the two opposing views represent, on the one hand, the opinions of those who see evolution as a slow process and the only one which shapes the destiny of the lyrebird in regard to the factor under consideration and, on the other, the views of those who consider the lyrebird capable of learning from experience and to have the capacity of applying such acquired knowledge in its fight for survival.

There is much evidence to support the view that lyrebirds do have a great capacity for learning, especially with regard to survival. For example, when the chick leaves the nest, it is totally dependent on its mother for food and protection; yet, under guidance, at the end of a year the young lyrebird will have learned how and where to seek its food, where to roost at night and perhaps other things. As it begins its life of self-sufficiency, it acquires further knowledge of its immediate environment. In due course the young lyrebird, through constant exposure to learning situations and assiduous practice, becomes proficient at singing and displaying on mounds, logs, stumps and in trees, etc., but the initial efforts of a one-year-old lyrebird give little encouragement to the human observer to believe that the bird will ever be the master of both of these functions, through a combination of innate and acquired behaviour.

Lyrebirds have remarkable memories. Like other birds, they have a capacity for 'absorbing' the physical details of their environment, so that they can move quickly from one part to another without losing themselves. Lyrebirds have favourite perches and drinking places and they can move unerringly from wherever they happen to be in the forest to a particular perch or drinking place. They remember not only the route, but also the physical thing which they want at any time. I used to spend much time with a particular bird (Spotty) and was intrigued by the manner in which he would suddenly cease whatever he was doing at the time (e.g., feeding or preening) and lead me down the hill to the base of a particular olearia tree, where he knew that he could always find food.

Other examples could be given of the lyrebird's capacity for learning and remembering things which conferred some benefit; but let us return to the question of whether a female lyrebird could be influenced by the recollection of some past experience when selecting a site for a nest. In 1938 a lyrebird built at ground level on the northern slope of Sherbrooke Gully, about 200 m from the Falls. When the chick was about four and a half weeks old, it was taken by a predator. In the following year, the bird built about two metres above the ground in an old stump and, in succeeding years, in similar situations, but not again at ground level.

Of course, this one example does not prove that, when the lyrebird was beginning to build in 1939, she remembered her unhappy experience of 1938, but the possibility that a particular lyrebird *could* remember and act accordingly should not, I think, be dismissed. So, while it is clear that the lyrebird species as a whole has not yet adopted, as part of its evolutionary pattern, the practice of building above ground level as a defence against predators, there may be individual cases in which this occurs. We need extensive data on the range of heights of the lyrebird's nest before the fox was introduced, and this obviously cannot be obtained. In the meantime, Reilly's data indicate that, in Sherbrooke Forest, some 40 per cent of nesting females built clear of the ground, but the reason why some lyrebirds build at heights of 20–30 m above the ground is not apparent.

In suggesting that a lyrebird might be motivated by unhappy memories when selecting a nest site, I was not implying that the bird had developed a capacity for reasoning, that is, of recognizing the relationship between cause and effect. If she had, it might be reasonable to assume that she would realize that, in building a high nest, she had trapped herself into a situation which was going to impose great strain on her during the next thirteen to fourteen weeks, at least; because, every time she had occasion to visit the nest, she would have to work her way laboriously from the ground, by springing from branch to branch in

*A female lyrebird pauses before springing across to feed her chick.*

her upward flight, requiring two minutes or longer. Of course if this were her first experience she would not have the opportunity of recalling these difficulties at the time. But there is another hazard: the sight of those luscious worms wriggling in the lyrebird's beak often proves too great a temptation for the ubiquitous kookaburra which frequently attacks her in flight, causing her to spill her hard-won food-parcel. I saw this happen several times in Sherbrooke in 1941 while I was watching the lyrebirds carrying food to two high nests. Lyrebirds, in their mimicry, often share a moment of laughter with the kookaburra, but there is no love lost between the two species, and it is always the kookaburra which is the aggressor.

I have already referred to one of Sherbrooke's celebrated lyrebirds, Droopy, which was so named because she had a damaged wing. There was an immature male with a damaged wing which I often saw during 1948–49, and it seems very likely that both birds suffered their disabilities in the course of their descent from high nests when they were inexperienced fledglings.

There has been some controversy regarding the reuse of the previous season's nest. Campbell (1884) could find no evidence of lyrebirds' having used their previous season's nests, although they 'apparently built on top of their old homes', while Jackson (1907) frequently observed that the birds built on the sites of their old nests 'year after year, if they had not been tampered with or the egg removed'. Hindwood (1955) gave details of a particular nest or nesting site which was used three times (1938, 1939 and 1942) in five years, and of another which he thought was used every second year from 1949 to 1955. Only once have I found the same nest used in successive years, the bird refurbishing her well-preserved 1941 nest in Sherbrooke Gully for a successful occupation in 1942. In another case the same site was used in 1943 and in 1945, while another bird used a site in 1971 and again in 1974. On 25 August 1968 the late Fred Waller, who was at that time the ranger at Glenaladale National Park, wrote to me as follows: 'Three weeks ago, I noticed the old lyrebird's nest opposite the iron railing had been pulled down. Today I saw a bird putting the finishing touches on a completely new nest. If she goes on and uses it, then it would be three times in four years that the same site has been used.'

I am happy to say that she did 'go on' and I was able to photograph the female flying away from the nest, after having fed the chick, on 17 November. The nest was on a cliff face, about five metres above the creek bed.

Roberts (1922) observed that Prince Edward's lyrebird (see chapter 11) reuses old nest sites, the old nest being allowed to settle down for a few years to develop a solid base on which to build the new nest.

# Time of nest-building and laying of egg

Some females spread their nest-building over several months, while others proceed more rapidly. I once saw a female in Odell's Gully, Sherbrooke Forest, gathering sticks in February, but she seemed to be engaged in some form of distraction behaviour rather than in serious nest-building. However, Coyle (see page 104) reported that his captive female once built a nest in February, and other females have been seen building nests in March. One such nest, discovered at the platform stage, was complete at the end of August and contained an egg on 8 September. In another case, a well-known female had completed her nest before the end of April 1939, but did not lay her egg until early in July. Another bird was observed building on a high bank above a popular walking track in Sherbrooke Forest in mid-April 1956. She had scratched out a niche in the slope and was transporting sticks to the site, but she attracted too much public interest and abandoned the nest. She resumed building in a more secluded site, about 30 m away, and duly produced a chick at the end of August. In another case, a nest was built in Sherbrooke Gully in July 1948 in about twenty-two to twenty-four days. Bradford (1933) found the beginnings of a nest in the Sydney district on 11 June; the nest was complete on 18 June and the egg was laid on about 2 July. Thus the period from the beginning of construction to the laying of the egg was about twenty-one days. Roberts (1922) observed that *Menura edwardi* (see chapter 11) required about two months to complete a nest, the egg being laid in July.

The earliest record of egg-laying of which I am aware is that provided by J.C. Mahan (1905) who found nests in the Woods Point district in Gippsland on 6 and 9 May, each containing 'the usual egg'. Hindwood (1955) found a nest near Sydney containing an egg before the middle of May. The latest egg-laying known to me was that of a Sherbrooke lyrebird which laid her egg during the first week in September, the chick leaving the nest early in December (Reilly, 1970).

# The egg

In size and shape the lyrebird's egg resembles that of the domestic hen, but there is considerable variation in shape, dimensions and colour. I have examined the H.L. White Collection at the National Museum, Melbourne, and what follows is a summary of the data pertaining to those eggs. The original designations have been retained for ease of reference. Eggs of *Menura novae-hollandiae*, collected from different parts of New South Wales, ranged in length from 57.7 to 71.5 mm and in maximum diameter from 41.2 to 47.1 mm, the averages (n = 16) being 62.5 and 44.2 mm. Most eggs have the usual shape; that is, a smaller, more tapered end and a larger, more rounded end; however, some remarkable variations were observed. In one case, the egg was an elongated ellipsoid with two sharply tapering ends; in another case, the egg had 'two large rounded ends'.

Eggs of *'Menura victoriae'* in the Collection ranged in length from 60.8 to 66.3 mm and in maximum diameter from 42.4 to 46.5 mm, the averages (n = 8) being 63.5 and 44.0 mm respectively.

Only three eggs of *'Menura edwardi'* (see chapter 11) were available, ranging in length from 57.9 to 60.5 mm and in maximum diameter from 41.4 to 43.4 mm, the averages being 59.5 and 42.4 mm respectively.

The colour of the lyrebird's egg shows wide variations both throughout the range and within a particular district. The eggs comprising the Collection vary in colour from very pale, almost white, through light stone-grey to darker grey and greyish-brown to very dark brownish-black, with darker splotches, especially at the larger end. Eggs from the Sherbrooke district usually have a slatey-grey ground colour, with darker splotches (mainly at the larger end); the ground colour varies from light to dark, and some eggs are light-to-dark biscuit-brown in colour.

The Sherbrooke Survey Group found that, in Sherbrooke Forest, over 80 per cent of the eggs were laid during the period from the second half of June to the end of July, the latest laying being at about the end of the first week in September.

Because eggs lose weight during incubation, unless one is fortunate enough to be able to weigh the egg shortly after it has been laid, values obtained under field conditions will probably not be the true weight of the fresh egg. Over the years I have weighed a number of eggs, usually after having found them addled, after the female had deserted the nest. In 1947 two such eggs weighed 50 g and 64 g respectively. In 1974, an egg weighed 50 g about ten days after having been laid and produced a chick which weighed 40 g at the age of two days. In another case, an egg weighed 56 g about ten days after being laid and 48 g thirty-five days later, when the female deserted because the egg failed to hatch. The shell of blown eggs weighed 3.7 g (range: 3.3–4.3 g; n = 3).

Lill (1979) gave a value of 62.9 g for the weight of a lyrebird's egg and 46 ± 6 g for the weight of a chick (n = 11). Therefore, assuming a value of 3.7 g for the weight of the shell, it would seem that the loss in weight during incubation is about 20 per cent, which is similar to that obtained for other species (Wallace, 1963).

Although the lyrebird normally lays only one egg each season, there are several records of two eggs having been found in the same nest. Somewhere 'in the wilds of Gippsland', on 24 July, H.C. Chandler found two such nests; one contained two fresh eggs while, in the other, 'incubation had begun'. In one case, the eggs were 'as like as two peas' (Campbell, 1900).

The two similar eggs *might* have been laid by the same female, but the origin of the others is conjectural. If, by 'incubation had begun', Campbell means that embryos were present, it follows that the eggs were fertile, but it is not clear whether they were laid by the same female or fertilized by the same male.

*A lyrebird taking a drink from the saucer-shaped top of a large fungus.*

*A mature male lyrebird displaying, facing slightly to the right, his body concealed beneath his tail feathers.*

*An immature male lyrebird, displaying to a fern, jumping from one foot to the other. Mature birds behave likewise.*

*At the conclusion of the display, the lyrebird raises his crest and walks off the mound. (This was Blue/White, Spotty's 1963 chick, towards the end of his eighth year.)*

*A mature male lyrebird and an immature male, engaged in a fight, with their claws wrapped around each other's legs, resulting in a stalemate.*

In another case, on 25 August 1893 D. LeSouef found a 'doublet'; but the eggs were different, one being lighter in colour and larger than the other, suggesting that two females might have been involved. Next day, LeSouef found another nest containing a fresh egg, which he took; but, about three weeks later, he flushed a female off the same nest which then contained another egg. This was slightly addled, and LeSouef thought that it had been laid shortly after his first visit to the nest.

Campbell gives yet another example of a second egg having been found in the same nest after the first had been taken. On 1 October 1892 Campbell, Chandler and LeSouef visited a nest from which Chandler had taken an egg two months previously; imagine their surprise when they saw a female fly out of the nest which then contained a fresh egg.

LeSouef seems to have had an uncanny knack of finding 'doublets' because, on 12 September 1900, he discovered a 'pair' in a nest near Gembrook, Victoria, which are now housed in the H.L. White Collection at the National Museum, Melbourne. These eggs are very dissimilar, one (62.3 x 46.2 mm) being yellowish mid-grey in colour, the other (60.7 x 45.7 mm) being light grey. These two eggs could have been laid by different females. In another instance, A.C. Stone (1916) took a partially incubated egg from a nest in South Gippsland and, thirty-five days later, he took another identical egg from the same nest, suggesting that both of these eggs could have been laid by the same female.

In more recent times, Hindwood (1960) described some remarkable experiences with 'two eggs in the same nest', in the Sydney area. On 9 May 1959 he found a nest ready for the egg and, a week later, he observed a female incubating. The egg was infertile and, about the middle of July, it was removed. Returning to the nest on 19 June 1960, Hindwood found a second egg in the nest, but its condition indicated that it had been laid a considerable time previously — probably shortly after the removal of the first egg. On the same day, he found another nest about half a mile from the 1959 nest; the female was sitting on an egg which later proved to be infertile and was removed. On 27 August 1960 Hindwood revisited the area, but found the 1960 nest deserted. However, he then returned to the 1959 nest, from which he flushed the female which was sitting on an egg, which also proved to be infertile. Hindwood did not see a male lyrebird in the area, which no doubt explains why the eggs were infertile; but, although Hindwood was careful to point out that there was no proof that the same female was concerned in these activities (because she was not banded), there must be a strong presumption that this was so. It seems remarkable that the female had not sought a singing male during the breeding season, but it is possible that the rugged nature of the country reduced the carrying power of the voices of males from adjoining territories. Of course, the female's reproductive system might have been defective.

It appears therefore that, on rare occasions, 'multiple laying' may occur with lyrebirds, although it is perhaps stretching the point to describe the phenomenon as such. It seems hardly necessary to mention that the laying of an egg is not an isolated event in the life of a lyrebird, but is the culmination of a series of complex processes which occur in the bird (chapter 5), and the laying of a second egg, even after an interval, implies that the hormonal and related processes have again been triggered into action without, as Hindwood so aptly put it, the inspiration of the male.

The lyrebird does not begin to incubate until several days after laying the egg. Roberts (1922) and Chisholm (1960) remarked on this behaviour, which is so different from that of other birds, most of which begin incubation as soon as the last egg has been laid and some (e.g. owls) even earlier, that it calls for an explanation. Briefly, fertilization of the ovum results in the formation of a new cell which contains all the information necessary for the production of every organ, every tissue and every chemical which will be required to produce the ultimate bird. This cell divides into two identical cells each of which divides into two more identical cells and, by the time the egg is laid, the embryo is a small flat disc lying on the surface of the yolk and adhering to the inside of the vitelline membrane. The embryo may remain in this condition for a week or more without suffering; but, if not warmed up, it will eventually die. For the development to continue, the egg must be incubated and it seems likely that this begins when the pro-lactin level in the bird's blood reaches the required threshold (chapter 5).

Presumably, the female occupies the nest at night, but it is several days, possibly a week, before daylight incubation begins. Even then, she does not sit continuously. Probably every student of the lyrebird has experienced that feeling of utter frustration which comes when the egg has been found stone cold and the female nowhere in sight.

This aspect has been studied by Dr Alan Lill, of Monash University, who recorded the times spent by the female in the nest during the daylight hours and measured the temperature of the egg by implanting a minute temperature-measuring device within the egg.

Dr Lill found that, up to about eighteen days, during the incomplete incubation stage, females usually left the nest, after nocturnal incubation, within two hours of dawn and did not return until mid-afternoon or a little later. On an average, the recess amounted to about seven hours. Thus, during the daylight hours, eggs were commonly incubated for about five hours, the temperature in the immediate vicinity of the nest during this period, in Dr Lill's studies, being 8–10°C. In other parts of the lyrebird's range, temperatures could be much lower and it is possible that, in such cases, the nest attentiveness during daylight hours could be higher, but information on this aspect is lacking.

After about eighteen days, a more complete incubation regime was established. Usually, there were two recesses daily; morning recess began between dawn and 9.30 and lasted about six and a half hours, until early afternoon. The second recess generally began at about 15.00–16.00 hours and lasted about 30–90 minutes. Thus the total incubation during daylight hours was still about the same as previously, but the periods of absence were broken, so that the average temperature of the egg during this period was higher.

In the first phase of incubation, during the recess, the temperature of the egg fell slowly from about 29°C to within three degrees of ambient temperature, in about three and a half hours; but, when the bird resumed incubation, the temperature rose fairly sharply to 29–30°C. (The temperature during normal incubation is probably higher than the 29–30°C recorded by Lill, because it has been shown that in the case of the pheasant, if the egg is prevented from being turned, the temperature within the egg may vary from 39.5°C at the brood patch to 25°C at the bottom [Westerkov (1956)]. Of course, in practice, the birds turn the egg or eggs frequently to minimize temperature fluctuations.) Heating-up rates were about three and a half times cooling-down rates.

Because of the shorter average duration of the recesses during the second stage of incubation, the temperature of the embryo was maintained at a higher level than during the first stage.

Towards the end of the incubation period recesses were usually short, so that the average temperature of the embryo would be higher. Precise data on this aspect are not available, but it is clear that, during the first stage of incubation, the egg is exposed to considerable fluctuations in temperature which are reduced during the second stage and still further during the third and final stage.

It is now easier to understand why the lyrebird chick under natural conditions takes about forty-five days to hatch, whereas when incubated almost continuously by a domestic hen, it hatched in about twenty-eight days (Ward, 1940).

# 5 | The lyrebird's breeding cycle

One of the wonders of the avian world is the reproductive cycle which reaches a climax at about the same time each year, for a particular species, when courtship, nest-building, egg-laying and production of broods proceed in an orderly manner. For most species reproduction occurs in the spring, but there are notable exceptions, the lyrebird being in this category.

Much work has been done to elucidate the factors involved in the reproductive cycle and the interaction of such factors in particular cases. Numerous species of birds have been studied including the European robin, the chaffinch, the bower-bird, domestic fowl, ducks, starlings, etc. Briefly, the results indicate that it is a combination of external factors which, acting in unison with the internal reproductive system, leads to the seasonal maturation of the sex organs and makes it possible for mating and reproductive behaviour to occur at the appropriate breeding season.

External factors include the hours of daylight (increasing in the spring), rainfall, temperature, etc., but none seems more important than the availability of food for the parent birds and the chicks. Internal factors are the bird's neural and endocrine systems.

Birds, like most other organisms, possess a daily or *circadian* rhythm, which manifests itself through the fairly stereotyped behaviour, over a roughly 24-hour cycle, in regard to such aspects as locomotion, body temperature, metabolism, hormone flow, etc. The circadian cycle plays an important part in determining the appropriate time of the year for reproduction. The factors involved are very complex; for example, in the case of the lyrebird, the decreasing hours of daylight are not the sole determinant because, while many eggs are laid before the winter solstice, even as early as the first week in May, many also are laid after the solstice, even as late as September. The details and mechanisms of the processes are stereotyped, but the timing varies, possibly due to genetic and environmental factors.

The following brief summary is based on the work of Marshall (1954), Witschi (1960), Dorst (1973), Welty (1982) and others. The highly developed sense organs of the bird enable it to appreciate significant changes in the environment. These effects impinge on the central nervous system (CNS) of the bird; individual members of a given species may vary in minor degree in their responses to external stimuli and in their capacity to learn from experience, whereby they enhance their chances of survival, but they lack the ability to reason, that is, to recognize the relationship between cause and effect. In response to stimuli from the environment, the CNS initiates further nervous and endocrine reactions, causing the bird to react in accordance with innate behaviour patterns traditional to the species and to a given situation.

There is strong evidence to suggest that the pineal gland, situated in the fore-part of the brain, plays a major role in receiving and interpreting environmental signals and that it triggers the hypothalamus and pituitary into action at the appropriate time (Welty, 1982). In turn, the hypothalamus releases a stimulating chemical messenger (hormone) which, travelling down the hypophysial portal system, activates the pars distalis of the pituitary gland. This organ then liberates hormones which flow through the blood to the reproductive organs of the bird, male or female. The pituitary gonadotrophins stimulate the sex organs, resulting in the liberation of male or female sex hormones which, in turn, cause the seasonal developments of other organs which are necessary in reproduction. These hormones call into operation behaviour patterns concerned with combat, sexual display, the discovery of a mate, nest-building and, finally, coition. Such activities proceed in a more or less orderly manner, according to the state of the weather, food supply and other vital environmental factors.

Following the breeding season, there is a period during which the reproductive apparatus of the male remains unresponsive to further stimuli. In effect, for a particular hormone to produce its traditional effect on the target organ, the concentration of that hormone in the blood needs to be at a certain threshold. If it is below this, a bird may, in particular circumstances, attempt to engage in certain courtship activities, but will be unable to achieve the normal result. This refractory phase lasts until late summer–autumn, or even longer. Thus, there is an internal brake on breeding.

Although, in the case of the lyrebird, there is a lack of experimental data on which to formulate ideas concerning the factors underlying the events which occur, it is clear that there must be an integrated plan governing the lyrebird's reproductive cycle.

During the spring, the sex glands are quiescent and remain so until the late summer or autumn when, in response to a stimulus from the CNS, the hypothalamus triggers the pituitary which releases the follicle-stimulating hormone (FSH) which activates the ovary and testes of the female and male respectively. In the case of the male, the hormone is known as the 'interstitial-cell stimulating hormone' (ICSH), but it is the same hormone. Both male and female lyrebirds must be stimulated to embark on a programme of interrelated activities designed to ensure that subsequent events will follow in the traditional manner. The two cases will be considered separately.

# The female lyrebird

At some time during the autumn (sometimes later), the female lyrebird, in response to an internal signal (though the initial stimulus may have come from the CNS–hypothalamus–pituitary system) will become involved in the new breeding cycle. First, she must select a nest site; then she must commit herself to the task of selecting and gathering suitable nest-building material, transporting it to the nest site (which may be at a considerable height or otherwise difficult of access) and, finally, she must actually build the nest. The period which elapses between the beginning of nest-building and the laying of the egg varies from one bird to another and is subject to seasonal influences.

The signal mentioned above activates the ovary, which has the dual function of producing the female sex hormones and progesterone. Presumably, the various stages of nest-building are related to different (increasing) levels of female hormone, involving also, perhaps, prolactin (see below), and the delays which occur between the commencement of nest-building and the laying of the egg are due to temporary arrests in ovarian activity. However, at some stage, the chain of events is resumed and the female undergoes changes designed to produce a fertile egg. The ovary continues to increase in activity until a peak is reached in June–July (sometimes later; very occasionally, earlier), when there is a ripe ovum ready to be released from the follicle. In response to a signal from the ovary, the pituitary arrests the flow of FSH and releases another hormone, luteinizing hormone (LH), which causes the follicle to release the ripe ovum. If this is to be fertilized, coition must occur before the ovum has become involved in the complex chain of events as it moves along the oviduct collecting various membranes, albumen (egg-white), water, minerals and, finally, the shell, each of which is added at the appropriate stage through the mediation of special hormones released through the lining of the oviduct. Because the lyrebird normally lays only one egg in each cycle, it is essential that the stimulus which caused the ripe ovum to be released be arrested. It would be decidedly disadvantageous if the female were to lay two eggs at about the same time. To prevent further ovulation, the ovary secretes progesterone. Finally, the egg is laid; but, several days before this, in order to ensure that the egg when laid will be adequately insulated against the cold, the pituitary, stimulated by the hypothalamus, releases a hormone prolactin which causes the female to remove feathers from her flanks and lower abdomen, both of which are generously endowed with suitable feathers for this purpose, and to place them in the nest. There is no obvious brood patch, but the female meticulously ensures good body contact with the egg, which is enveloped in her feathers.

The egg having been laid, it becomes necessary for the female to be induced to incubate it; this is achieved through the mediation of prolactin. The hiatus between the laying of the egg and the commencement of incubation probably represents the period during which the prolactin level in the blood is being raised to the threshold which marks the beginning of incubation. Thereafter, the prolactin is maintained at the appropriate level to ensure complete incubation, probably through the tactile stimulus provided to the CNS by the pressure of the egg against the bird's body. This keeps the hypothalamus–pituitary system active, even to the extent of persuading a female lyrebird to continue to sit for a period of nine weeks, two weeks longer than usual.

What causes the female ultimately to abandon the nest (egg) is somewhat uncertain, but perhaps there is some innate factor which is triggered by the CNS when the nature of the tactile stimulus received through the abdomen changes as the egg is replaced by the chick. If this stimulus is not received by the CNS at about the correct time, the prolactin level falls and with it the urge to incubate.

If a chick is produced, another set of conditions begins to operate. A suitable prolactin level (resulting probably from the tactile stimulus received by the CNS–hypothalamus–pituitary chain) causes the female to brood the chick. The first brooding session must be long enough to enable the chick to grow its first coat of dark dorsal down, so that it will be adequately insulated against the cold and thereby enable the female to take a short recess from brooding to gather food. Presumably, the prolactin level is maintained to ensure the completion of the brooding regime (including attendant functions such as nest hygiene) and that the female will remain attentive to the growing chick after it leaves the nest, until it becomes self-sufficient at the commencement of the next breeding cycle. It seems likely that the tactile stimulus received by the female's CNS during the incubation and brooding phases is replaced by an auditory signal in the form of the chick's continual wheezings as the two birds move about the forest after the chick leaves the nest. Presumably, when the chick no longer needs the support of its mother, the pituitary ceases to produce prolactin and begins to stimulate the ovary again; or it may be that the seasonal conditions of the autumn stimulate the CNS, which shifts the hypothalamus–pituitary into gear, resulting in an increase in ovarian activity, which, in turn signals the pituitary to cut off the supply of prolactin. Whatever the mechanism, a new breeding cycle begins as the female–chick bond is severed.

As mentioned previously, the foregoing lacks experimental support, but the work of Hohn (1950), Hinde (1962) and Frieden (1976) suggests that prolactin, or some similar hormone, plays an important part in the activities referred to above. However, Eisner (1960) has questioned the dominant role suggested for prolactin. The difficulty of making hormonal assays on blood samples of the lyrebird can be appreciated because, whereas it appears to be routinely practical to sacrifice perhaps fifty to a hundred starlings or pigeons in an experiment on the role of, say, the thyroid gland in the behaviour of these species (Voitkevich, 1966), this is not the case with the lyrebird.

# The male lyrebird

With the release of FSH (ICSH) by the pituitary gland in the autumn, the testes of the male lyrebird become active again. The function of the testes is to produce mature sperm for the fertilization of the female ovum at the appropriate time and to produce the male sex hormone testosterone. Like the female, the male bird has a number of preliminary duties to perform before he reaches the peak of his sexual activity which leads to copulation. Although he will have been singing and displaying for several months before the onset of the breeding season, when the latter really begins, probably stimulated by increasing levels of the male sex hormone, he begins to show greater interest in the refurbishing of old mounds and preparation of new ones, and in singing and displaying. By early May, he begins to spend more time in courtship activities, in preparation for the final act of coition. As autumn gives way to winter, through the mediation of the ICSH and LH, his sperm will have matured and be ready for the female and, usually early in June, he is at his peak of sexual activity.

The copulatory act is preceded by a bout of intensely active display and song in which the male, followed by the female, moves from one mound to another until they are both ready for the final act. As a prelude to this, the male goes to a mound where he engages in full-face display and full song, while the female remains a little distance away, watching and listening. Suddenly, the male ceases the full song and brings his tail feathers almost together over his head, almost touching the ground with the feathers which he vibrates rapidly while uttering a continuous 'Blick blick blick blick...' note. This is the 'invitation display' and, when the female slowly comes on to the mound, he raises his tail to enable her to come closer to him and the two birds circle one another several times. Usually, their beaks touch for an instant. Suddenly the female is stationary and facing away from the male, in a crouching position, signal-

*The lyrebird uses his two large claws to remove vegetation to prepare a clear area in which to construct a display mound.*

ling that she is ready to accept him. She moves her tail to the side and the male mounts her, with his tail feathers closely packed and extended over his head, at an angle. He works his wings vigorously as he strives to bend his body downwards in order to make cloacal contact. The ridges of the caudal muscles are bright pink as the blood pulsates; the whole performance is brief, lasting perhaps ten seconds. Then the male dismounts, moves off the mound backwards, whilst still holding the tail feathers closely bunched and vibrating rapidly. Eventually, he will encounter some obstacle which will cause him to turn and, probably, run to another mound where he will display and sing strongly. Meanwhile, the female, after a few moments' rest, will settle her feathers into place and quietly leave the mound.

Obviously, once the ovum has been released from the follicle, it must be fertilized as quickly as possible. Multiple copulations enhance the ovum's prospects of being fertilized and, in June 1976, I met another observer in Sherbrooke Forest who informed me that, on that particular day, he had observed the birds in that area copulating no fewer than eight times. However, Ina Watson (1965), in her account of the mating of the superb lyrebird, recorded only one copulation in the course of two days and observed that, after copulation, the female appeared not to respond to the male's repeated invitations to join him on the mound again. However, it is possible that mating had occurred before the one act observed.

It is not known how long it takes for the lyrebird's egg to be laid after the ovum has been released from the follicle; in the case of the domestic hen, the period is about twenty-three hours (Harrison, 1964).

Presumably, the male lyrebird remains in a sexually active condition for several weeks after reaching his peak, to ensure that any females which reach their peaks of sexual activity later than normal may be assured of finding an active male; but, by August, most males in Sherbrooke Forest show declining sexual activity and many are in tail-moult by the end of August or early September. The sexual apparatus remains inactive throughout the spring and summer; but, in response to stimulation by the CNS–hypothalamus–pituitary systems, begins to develop again in the following autumn and a new breeding cycle is set in train.

# The young lyrebird

## Life in the nest

When the lyrebird chick emerges from the egg, at the end of its seven-week incubation period, it is blind and helpless, and weighs about 40 g. It is practically naked, having only a few hair-like feathers on its head and back, accompanied sometimes by a little black down. If left alone in this condition, in the low temperatures which prevail throughout the lyrebird's range, it would soon perish. But just before the chick hatches, the abdominal muscles contract and the residual egg-yolk is drawn into the system, thus providing nourishment during those first few critical hours (Bellairs, 1960; Harrison, 1964). This relieves the female of the need to feed the chick, so that she is free to sit on it and maintain a temperature of about 35°C. By the end of the first day, the chick is covered on the head and back with sooty black down. The underside is bare, but is well protected against the cold by a warm bed of feathers placed on the floor of the nest by the female for this purpose.

I do not know just when the chick receives its first meal from the female, but it is probably not until the second or third day. It is known that the domestic chick does not require food until two days after hatching. On one occasion, in 1974, I arrived at the nest just as the chick was breaking out of the egg, at 7.30 a.m. A few hours later, it weighed 40 g. Another chick, which I judged from the condition of its feathers to be about two days old, also weighed 40 g.

In preparation for the life it must lead after it leaves the nest, the chick develops its various organs to the point from which it can successfully embark on the next phase, and the following account is based on observations and measurements made on two chicks which were studied in Sherbrooke Forest in 1974.

## The eye

It appears that lyrebird chicks vary a little in the age at which they open their eyes. Reilly (1970) observed that the eyes open at five days, but Chick No. 1 had its eyes barely open at six days, while Chick No. 2 had its eyes firmly closed at nine days, but open at fourteen days. At forty-four days, the eyes, oval in shape, measured 12 x 10mm; they were dark brownish-black with blue irides. The eye of the mature male is likewise oval in shape, measuring 14 x 12mm, and is blackish with black irides.

## Feather development

To appreciate the significance of the various phases of feather development which occur, it is necessary to recognize that there is a plan underlying the processes involved and their sequences. No doubt down is grown first, that is, before the definitive feathers, because this imposes less strain on the organism and provides the necessary insulation more quickly. It is well known that an increase in temperature of 10°C approximately doubles the speed of a chemical reaction; thus, continuous brooding for the first day, whereby the chick's body temperature is maintained at about 35°C, results in at least a six-fold increase in the speed at which the chick acquires that first coat of

down, assuming an ambient temperature of 10°C. In fact, when I measured the temperature within the nest cavity, it was nearer to 5°C; however, the temperature under the chick's body (belly) was about 35°C. To measure temperatures, I was obliged to use a mercury-in-glass thermometer and it was noticeable that, during the first twenty-one days (that is, while the chick's belly was bare), the mercury rose rapidly to the maximum reading, because there was good contact between the subject and the thermometer bulb. After the belly had acquired a light cover of feathers, the thermometer took 7–10 minutes to reach its maximum reading (about 35°C), instead of less than three minutes, because of the insulating effect of the belly feathers. This demonstrates the high insulating power of the bird's feathers and why the first feather coat appeared on the dorsal parts of the chick, the ventral parts being well protected by the nest feathers. Later, the belly feathers began to grow in preparation for life outside the nest.

Dr Alan Lill has studied the time spent by the female in the nest, incubating the egg and brooding the chick. By measuring the rate at which the chick's body temperature fell after the female left the nest, he found that there was a gradual improvement in thermo-regulation up to ten days, after which the chick was essentially endothermic, and the female no longer brooded it during the greater part of the daylight hours. When the air temperature in the vicinity of the nest was 2–12°C, the temperature of the nestling fell 6–9°C after 30 minutes' exposure (that is, after the female ceased brooding) during the first four days; but, on days 9 and 10, the temperature fell only 2°C after 60 minutes' exposure. By this time, the nestling has acquired a good coating of dorsal down, which reduces heat losses, thus relieving the female of the day-brooding function and allowing more time for foraging. However, I have noticed that some females will enter the nest and brood the nestling, even after the tenth day. Thereafter, by virtue of the insulation and 'regular' feeding, the chick maintains its body temperature at an adequate level. The female

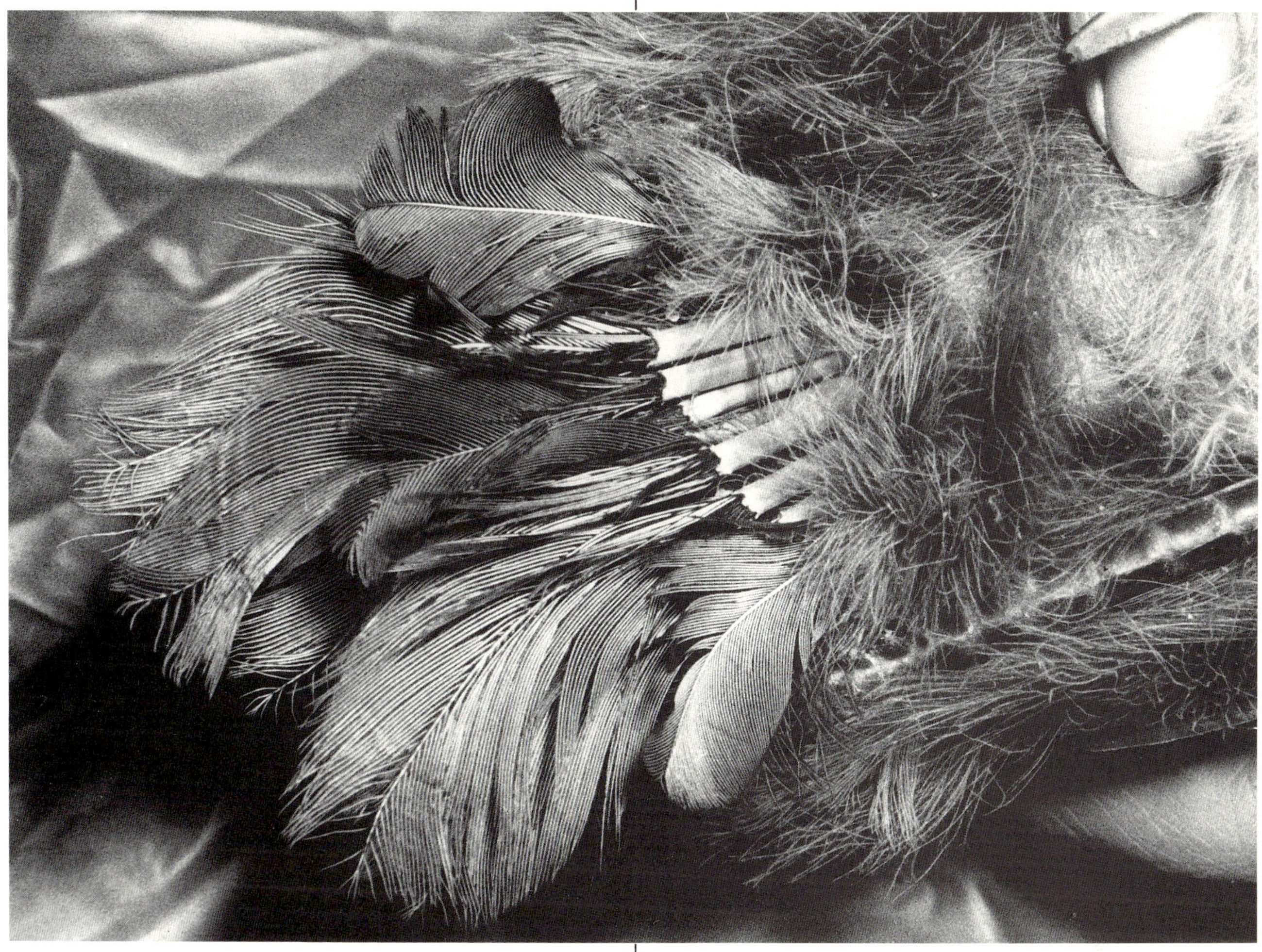

The underside of the tail of a lyrebird chick aged 42 days, just before fledging, showing that the tail feathers are still growing.

broods the chick overnight and I have observed this to occur until the chick is at least thirty-three days old, but I do not know just when overnight brooding ceases.

The upper parts, including the wings, develop quite rapidly, while the lower parts acquire feathers more slowly; but, by the time the chick is ready to leave the nest, it is adequately covered to enable it to undertake the next step in the species survival plan.

In addition to the sooty black down which covered the upper parts and wings during the first six hours, faint lines were also visible along the sides of the body below the wings, indicating the location of future feather tracts. At two days, the downy feathers showed further growth and, at six days, there was considerable development of the dorsal tract. At nine days, the legs and underside of the body were still practically bare, apart from two ventral tracts which were just beginning to grow. At thirteen days, the neck was still largely bare, but the back had a good cover of longish tract feathers which were pushing out the down, and the feather growth had extended to the upper belly and breast. The lower belly was bare, but faint streaks of emerging feathers were visible. These feathers were not preceded by the 'gollywog' down which characterizes the first covering of the newly hatched embryo. At twenty-one days, the body covering had extended still further and the belly had a thin covering of light-grey feathers. This general pattern of feather development continued, but the neck and face remained bare until towards the end of the fifth week. At forty-four days, Chick No. 1 had a good covering of rufous-coloured feathers on the throat and forehead, and the neck was also covered. However, at thirty-nine days, Chick No. 2 had a practically bare neck and only faint rufous colour on the forehead and throat, the breast being well covered with very dark feathers. At this stage, in both cases, the body was well covered with feathers which had replaced the down, except for a few tufts of down which still adhered to the tips of the head and body feathers.

Tail feathers were not visible at fourteen days, but must have been close to erupting, because, at nineteen days, the visible part was 19 mm long in the case of Chick No. 1 and a little shorter in Chick No. 2. When the chick leaves the nest, all tail feathers are still in their sheaths and continue to grow for about six to eight weeks thereafter. Over the years, chicks have been observed to grow tail feathers up to about 140–160 mm long before leaving the nest.

## Wings

By the ninth day, the sheaths of the wings had grown to a length of about 25 mm and the feathers were just emerging; by the thirteenth day, growth was so rapid that it was noticeable in the course of a day's observation (seven hours). At the end of six weeks, the wings appeared complete, but there were still bare areas on the sides of the body beneath the wings and the wing feathers were still in their sheaths. The development of the wings of a chick found dead in the nest on 2 October 1971, aged about thirty-three days, demonstrates that the wings continue to grow for several weeks after the chick leaves the nest. This view is supported by the fact that a nestling specimen in the Australian Museum (Sydney) had tail feathers 160–170 mm long, which, along with the length of the digits, indicated that the chick was at least seven weeks old; the wing feathers were still in their sheaths. This appears to be the general rule, because I have observed that at the time of leaving the nest, the young of the grey thrush, white-browed babbler, yellow-tufted honeyeater and crested bell-bird all had their wing feathers still in their sheaths, i.e., they were still growing.

## Feeding of nestling

In Sherbrooke, the principal items which the female has been observed to feed to the nestling are worms, grubs and small crustaceans (*Talitris* sp.). Sometimes a centipede is included in the diet; but, although I have seen adult lyrebirds eating scorpions and land yabbies, I have not seen the female feed such items to the chick. However, Robinson and Frith (1982) have reported on a much more comprehensive diet for lyrebird chicks in Tidbinbilla Nature Reserve (ACT) and have confirmed my own observations that the female is not selective in her choice of food for the chick, taking whatever is available. Occasionally, the female finds a large worm and feeds it to the chick; the latter usually finds it impossible to swallow it and therefore sits in the nest with the unswallowed part projecting from its beak until the female returns and pushes it down its throat before delivering the next feed.

Continuous observations were made on Chick No. 2 at the age of twenty-eight days over a period of 200 minutes, during which

*A female lyrebird depositing the chick's dropping in the stream.*

nine feeds were given to the chick at intervals of 10, 15, 20, 40, 40, 30, 20 and 25 minutes (average, 25 minutes). One week later, over a period of 308 minutes, thirteen feeds were given, at intervals of 15, 35, 30, 25, 15, 25, 20, 15, 30 and 13 minutes (average, 25 minutes). A week later, seven feeds were given at intervals of 27, 33, 30, 20, 50 and 33 minutes (average, 29 minutes). Bad weather precluded observation over a long period of Chick No. 1; but, over short periods, the interval between feeds was 10–12 minutes. Frith and Robinson (1982) found that in Tidbinbilla the interval between feeds was 6–42 minutes, with an average of 19½ minutes.

## Nest sanitation

When the chick needs to defecate, encouraged by soft clucking noises from its mother, it turns round in the nest and backs towards the platform to deliver the dropping directly into the beak of the female, which either transports it to a stream (up to 100 m away) where she submerges it, or buries it in the ground some distance from the nest. The dropping consists of the excreta encapsulated in a tough gelatinous sac. The importance of keeping the nest clean and free from smells which might attract predators is obvious and the methods adopted by different species of birds have been described by Donald F. Thompson (1931). The method adopted by the lyrebird is used by many other species of birds. Portman (1960) states that the mucilagenous sac is formed by the colon only while the chick is in the nest. The need for such a process exists only at this time, to facilitate transportation and to ensure the cleanliness of the nest.

There is no doubt that both chick and parent bird are aware, however unconsciously, of the need to keep the nest clean, and their behaviour reflects this awareness. If the female lyrebird is absent from the nest longer than usual and the chick needs to defecate, it is quite apparent that the chick is reluctant to soil the nest. Under such circumstances, it shuffles around in the nest and moves backwards towards the front. It utters a shrill piping note which usually brings the mother running. But, occasionally, the chick is 'caught short' and delivers its dropping on to the platform, after which it settles down in its bed of feathers. Reilly (1970) reported that chicks have been found dead on the ground below the nest, having overbalanced when backing on to the platform to defecate. I have

seen this happen and have returned several chicks to the comfort of their warm nests. On returning to the nest, the female either transports the dropping to the stream or swallows it. If the dropping has been damaged, she meticulously picks up the pieces and swallows them. However, the dropping is not projected into the air (by the chick) and seized in flight, as proposed by Barruel (1973).

During the first period of continuous observation referred to on pp. 66–8, five droppings were taken from the chick at intervals of 45, 80, 30 and 45 minutes (average, 51 minutes); during the second period, seven droppings were taken at intervals of 45, 65, 75, 35, 30 and 45 minutes (average, 50 minutes); in the third period, four droppings were taken at intervals of 93, 30, 50 and 33 minutes (average, 52 minutes).

Droppings retrieved from the stream were seen to be roughly ellipsoidal in shape, or like a tadpole with a long tail. They were about 45–50 mm long and 25–30 mm wide. At one time, there were twenty droppings in the pool in the creek.

## Leaving the nest

A question which is often asked is, 'How do the chick and female know when the time has come for the chick to leave the nest?' I used to think that, at this time, the female withheld the chick's food, standing in front of the nest and making soft clucking noises to entice the chick to follow her into the forest. The 'life-at-large' must seem more attractive to the chick when its mother's beak is full of wriggling worms. This may still be a factor, but probably a secondary one. I have watched a chick, which seemed ready to leave, standing on the platform outside the nest for as long as two hours before moving off. In other cases, the departure has taken only a few minutes. It seems most likely that the provision of the internal sanitary service by the colon has a special significance for both chick and female, and the cessation of this service is the signal to both birds that the time has come for the chick to depart. Once the chick has left the nest, its

*A female lyrebird enticing her chick from the nest.*

faeces, like those of the adult birds, are semi-liquid.

A related question is, 'Does the chick return to the nest at night-time, after leaving?' Tregellas (1921) suggested that the presence of mud on the feet of a chick in the nest indicated that it had returned to the nest after having spent the day in the forest; but there may be another explanation. The female often collects mud on her feet while foraging and some of this could have dropped off on to the platform while she was feeding the chick. I have seen this happen. Some of this mud could have been collected by the chick when it was on the platform to defecate. The presence on the platform of a white gelatinous substance, as described by Reilly (1970), could have been due to a breakdown in the colon service, during the transition stage. The only chick which I ever saw back in the nest at night-time, after having left the nest in the morning, was one which I caught and returned to the nest for the first two nights after it had left. I did this because, at the time, I thought it was too young to be out of the nest! Nowadays, I leave such matters to the judgment of the birds themselves.


## Effect of temperature

One of the most remarkable aspects of the breeding behaviour of the lyrebird is that usually the egg is laid during the coldest months of the year (June–July–August).

Data were not available for temperatures in the areas used by the lyrebirds, but figures supplied by the Bureau of Meteorology for districts within the range of the superb lyrebird show that the average monthly minimum temperatures during June, July and August are often below 5°C, sometimes even below 0°C. On some days, temperatures are well below the average minimum values, in some districts, and it may safely be assumed that, at the nest sites, the temperatures are even lower. In many years, during the breeding season, snow has fallen at Sherbrooke, causing damage to some nests, one having been found crushed under the weight of snow.

Robinson and Frith (1982) recorded temperatures as low as 0°C in Tidbinbilla and Roberts (1922) found temperatures as low as −11°C (=12°F) in an area occupied by *Menura edwardi* in southern Queensland, during the breeding season.

The air temperature near the nest of Chick No. 2 and in the space above the chick in the nest was often 7°C, but the temperature under the chick was 35–36°C.

As already mentioned, Lill demonstrated that the incubation regime consists of three phases, namely, a phase of incomplete incubation extending over about eighteen days, followed by a second phase in which the daylight incubation is carried out in two 'shifts', with shorter recesses, thereby achieving a higher average body temperature, and a third phase, towards the end of the incubation period, of intensive incubation. On the whole, the 'attentive index', that is, the percentage of the average day-length incubated during the second phase, was about 45 per cent, as compared with 60–80 per cent for most other passerines.

The irregular incubation regime of the lyrebird prolongs the process. Incubation periods range from 45 ± 3 days (Reilly, 1970), and from 41–44 to 54–55 days (best mean estimate 47 days) (Lill, 1979). The longest incubation period of which I have personal knowledge was fifty days.

The unusual incubation behaviour of the lyrebird has long been known to naturalists, and over forty years ago an enterprising student of the lyrebird decided to settle the question of how long the chick would take to hatch if the egg were incubated continuously instead of intermittently. So John E. Ward (1940) removed a freshly laid lyrebird egg from a nest in the Burragorang Valley in New South Wales and placed it under a broody black Orpington hen at his farm, which was not far from the nest. In exchange for her egg, he gave the lyrebird a hard-boiled hen egg, stained to match the lyrebird egg. The hen rarely left her nest, except to feed and drink and, after twenty-eight days, Ward had the satisfaction of seeing the lyrebird chick

*A mature male lyrebird feeding his four-month-old chick whose mother had died. Photo by Kevin Mason.*

*Cradle of a lyrebird's nest, Sherbrooke Forest.*

*A lyrebird's nest in Sherbrooke Forest, built among the ferns at the base of a tree-fern.*

*Lyrebird eggs from different districts.* Clockwise from bottom, centre: *superb lyrebird, Ferntree Gully; Agnes River, Victoria; Mt Toole-be-wong, Victoria; South Gippsland; unkown; Clarence River, NSW; Harrietville, Victoria; Walcha, NSW.* Centre: *Prince Edward's lyrebird, Stanthorpe district, south-east Queensland.*

*A female lyrebird settling down on her egg in the nest.*

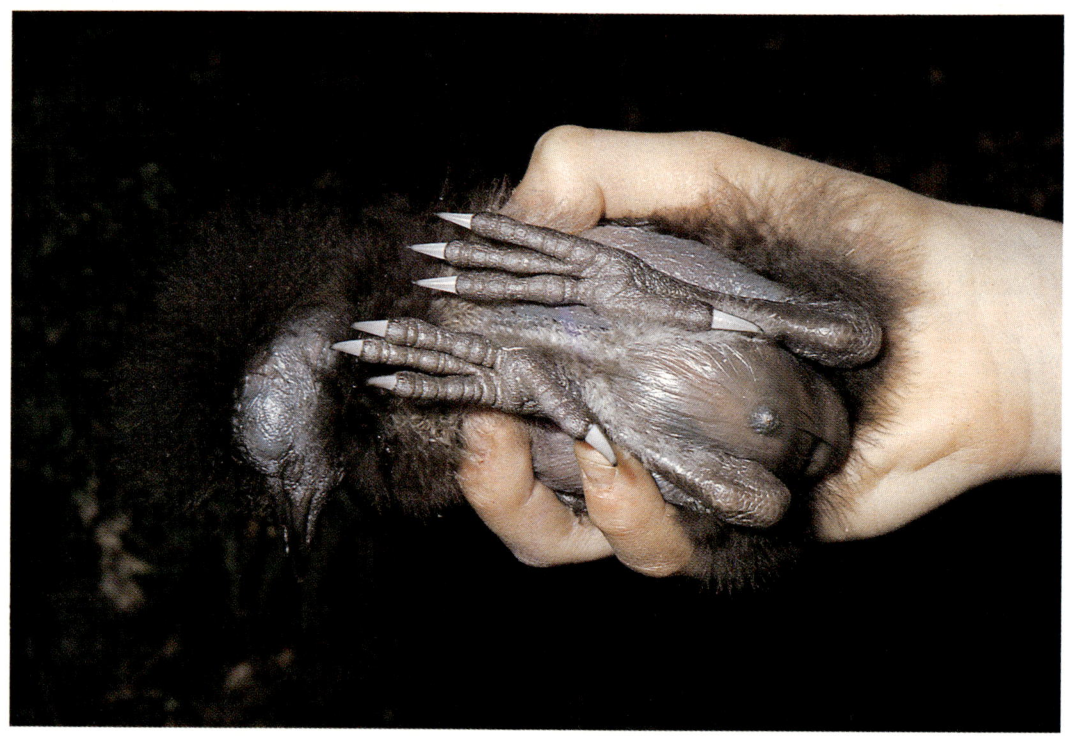

*A lyrebird chick aged nine days, showing the eyes closed, dorsal black down, bare underside, scar from absorption of residual yolk before hatching, and vestigial egg tooth.*

*A lyrebird chick aged forty-four days, showing the rufous coloration of the throat and frons feathers.*

shortly after it had hatched. Ward hastened back to the lyrebird's nest, where the female was still incubating the hen egg, and placed the lyrebird chick in its rightful nest, but later found that the female had rejected it, while she continued to sit on the hen egg which she incubated for nine weeks before deserting the nest. Had Ward removed the hen egg when he placed the lyrebird chick in its natural home, it is possible that its mother would have accepted it and reared it. No-one seems to have repeated Ward's experiment, but this should be worthwhile. The importance of temperature in relation to incubation period is clear.

Lill measured the body temperature of the chick by deep cloacal insertion of a thermocouple linked to an electronic thermometer and recorded values of 30°C or lower, except on the tenth day, when a temperature of about 33°C was recorded. My temperature measurements were made by placing a mercury-in-glass thermometer under the chick's belly or by holding it in the nest cavity above the chick. The thermometer was withdrawn from under the chick after a short period and read as quickly as possible. Some practice was required to ensure that the top of the mercury thread was picked up without delay; but, if the readings were slightly in error, it is clear that they would have been too low, rather than too high. Nevertheless, I recorded body temperatures of 35–36°C for a chick aged 14–44 days.

Sometimes one finds chicks dead in the nest, but otherwise unharmed. The most likely explanation is that the mother has been taken by a predator or unable to return to the nest for such a long time that the chick's body temperature fell below the required level.

The problems associated with the incubation of the egg and the brooding of the chick by a lyrebird, unaided by her mate, can only evoke the admiration of all those who have had experience in the raising of a brood. Rowley (1974) makes an interesting observation on the brooding behaviour of the Australian raven; both male and female birds cooperate in brooding activities, but the female is largely confined to the nest for the first three weeks, while the young birds are acquiring an adequate covering of feathers which maintain their body temperature at a safe level. This, I think, provides support for my views regarding the development of the feathers of the lyrebird chick, namely, that the protein devoted to feather formation must be deployed in the manner most advantageous to the species.

One other aspect of the lyrebird incubation process and its division into three phases invites the obvious question: what do the three stages signify? I am not aware of any research on the embryology of the lyrebird, but a brief account of some of the parallel developments in the domestic hen may help to explain the incubation phenomena.

Embryonic development consists mainly of two processes; firstly, an increase in the number of cells and, secondly, a continual change in the structure and arrangement of the cells. The first stage brings about an increase in size, while the second results in the differentiation of the various organs and a number of far-reaching changes in the biochemical and physiological activities of the developing embryo (Bellairs, 1960).

Work on the development of the embryo of the domestic hen has provided much valuable information which, it would seem, has application here. For example, when the egg is laid, the embryo consists of a small flat disc of cells (blastoderm) lying on the surface of the yolk and sticking to the vitelline membrane. The embryo may remain in this condition for a week or so; but will eventually die, if it is not warmed up. When the hen (or female lyrebird) begins to incubate, cell division proceeds in response to the rise in temperature, and the embryo increases in size; but, as yet, there is no differentiation of the cells to form the various organs, chemicals, etc., which go to make the chick. In the case of the domestic hen, this process begins about thirteen days after the laying of the egg and continues until the chick is ready to leave the egg.

The only contribution a sitting hen can make to the development of the embryo is to maintain the appropriate temperature. The domestic hen sits more continuously than the lyrebird and therefore maintains a higher average temperature, which in turn reduces the incubation period.

It seems likely that, during the first eighteen days of (incomplete) incubation of the lyrebird egg, the embryo is developing through the first stage. The longer period is to be expected, because of the lower average temperature in the case of the lyrebird. Once differentiation begins, it would seem reasonable to expect that a higher metabolic rate would be required to enable the embryo to continue to develop and this is achieved by maintaining a higher average temperature. This is provided by varying the incubation regime and, as the embryo develops further towards the final stage (hatching), the higher metabolic rate demands even higher average temperatures which are achieved by more intensive incubation, as in the third stage referred to above.

## Growth of the chick

Until comparatively recently, the only information on the weight of a lyrebird chick at various stages of development appears to be that of Campbell (1941). In 1974 I obtained further information (unpublished) on the two chicks referred to above, and Lill (1979) has since provided additional information. To facilitate comparisons, the weights recorded by Campbell and myself have been plotted against the age of the chick and weights at comparable ages are set out in the following table, along with Lill's data.

The rates of growth were approximately linear, but it is clear that lyrebird chicks vary somewhat in their age–weight relationships.

It is obviously difficult to obtain data for the chick after it has left the nest, but an eleven-month-old chick, which had been reared by hand at the Healesville Fauna Park, weighed 1080 g. Information on mature birds is rarely obtained; a female specimen at the National Museum, Melbourne, weighed 908 g, as did a mature male which came into my hands in 1969 (see chapter 9). Fleay (1952) recorded a figure of 1248 g for a mature male which was trapped and transported to Tasmania in 1949 (see chapter 10).

## Weights of lyrebird chicks at different ages

| Age (days) | Weight (g) | | | |
|---|---|---|---|---|
| | Campbell | Smith No. 1 | Smith No. 2 | Lill |
| 0 | | 40 | | |
| 2 | | | 40 | |
| 4 | 64 | | | |
| 10 | 85 | 150 | 145 | |
| 20 | 355 | 260 | 290 | |
| 30 | 500 | 375 | 440 | |
| 40 | 560 | 485 | 545 | |
| 42 | | | 545 | |
| 44 | 505 | | | |
| 47 | (average fledging age: n = 11) | | | 571 ± 63 |

In 1974, I measured the tarsometatarsus (tarsus), claw-span and middle front and rear toe-nails (digits) of chicks Nos. 1 and 2 at different stages of their development. To facilitate comparisons, graphs were prepared by plotting the respective dimensions against age and reading off the relevant values at stated ages. The results were as shown below.

# Life in the forest

When the chick leaves the nest, at the age of six to seven weeks, it is equipped with a pair of strong legs and large feet which provide balance as it follows its mother around the forest. At first, it is somewhat unstable, but is very active after a few days. On first leaving the nest, it is led by its mother to a place of concealment, such as a large log, a clump of ferns, the buttress of a large mountain ash, a clump of sword grass, or some similar situation. Within a few days, it leaves the ground in favour of an arboreal perch, but for several months remains dependent on its mother for most of its food. As the chick grows in strength and agility, it moves to higher perches, especially in olearia trees, possibly because the close proximity of the branches facilitates 'climbing', but perhaps also because of the protection afforded by the foliage. The chick frequently moves from one perch to another between feeds, causing the female to search for it. The female keeps the chick informed of her whereabouts by a soft clucking sound, while the chick, more especially when it is on the ground, responds with a soft wheezing sound. Within a month of leaving the nest, the chick is able to fly or glide quite swiftly and, at the end of the day, encouraged by the clucking of its mother, may be seen following her in a zig-zag upward course to the overnight roost.

Gradually, through close association with its mother, the chick learns to scratch among the forest litter, but appears slow to realize that the small creatures it uncovers are its food, and waits for its mother to feed it. However, it does eventually learn to recognize its food; but, even when four to five months old and frequently much later, the chick begs for food from its mother which invariably responds, even though the chick may be perched

# Length of tarsus, claw-span and digits of the superb lyrebird at different ages

| Age (days) | Tarsus (mm) | Claw-span (mm) | Digits (mm) | |
|---|---|---|---|---|
| | | | Middle front | Rear |
| 5 | NR | 25 | NR* | NR* |
| 10 | 53 | 48 | 6 | 8 |
| 20 | 80 | 84 | 11 | 10 |
| 30 | 113 | 120 | 16 | 20 |
| 40 | 121 | 130 | 20 | 22 |
| 44 | 123 | 140 | 22 | NR* |
| Mature male (museum specimens) | 114** | 140 | 29 | 38 |

*NR = not recorded.   ** Range: 100–125; N = 10.

*The author's seven-year-old daughter, Helen, enraptured by a lyrebird chick.*

in a tree, 10 m above the ground. In such situations, the chick sometimes behaves in a manner which suggests that it is either incredibly stupid or very determined, or perhaps that it is a slow learner in some respects. For example, I once observed a young lyrebird (five months old) perching in a dry tree, the branches of which were wet and slippery from overnight rain. It appeared that the young lyrebird had decided that it wanted to move to a higher perch, so it proceeded to *walk* along a branch which sloped upward at a sharp angle. The bird slipped backwards, but repeated the attempt half a dozen times. It could easily have opened its wings and sprung to its objective, but seemed determined to walk. As the chick continued to fail in its efforts to reach the higher perch, it squawked as if expressing its mounting frustration; but finally it paused, and then sprang across to another branch from which it sprang to the place it had been striving to reach. Actually, I suppose that it was practising the zig-zag upward 'flight' behaviour which appears to be traditional to the species.

I do not know just when the young lyrebird meets its male parent for the first time, but have often seen a chick aged four to five months in the company of both parents, sometimes along with the previous season's chick, which has competed vigorously with the younger bird for its mother's favours, though without success. The birds have remained together for one to two hours, scratching for food within a few feet of one another. Rawnsley (1863) reports having seen a complete lyrebird family — male, female and 'half-grown' chick — early in September, in the Illawarra district of New South Wales.

Even when the next breeding season is in progress, in May, June and July, and the female is involved in courtship activities, the chick is reluctant to leave its mother. At times it remains in close attendance, even when the female is attracted to the mound by a displaying male, and I have seen the nine-month-old chick, sometimes accompanied by the previous season's chick, follow the female to the edge of the mound when the female went on

to the mound with the male. Even when the female is involved in nest-building, the chick remains close by and, later, when the female is feeding the new chick in the nest, I have seen the yearling chick beg so persistently that the female has given it the food which she was carrying to the chick in the nest. Some females endeavour to break the bond with the chick by rushing at it, with obvious annoyance; but, within a few minutes, I have seen the persistent year-old chick returning to the vicinity of the nest to resume its begging. Sometimes, a young lyrebird, deserted by its mother, will beg from other young lyrebirds, but they are less tolerant and often give the beggar a solid push with their claws, sending it on its way. Eventually, however, the young bird becomes self-sufficient, and I have often seen a one-year-old lyrebird scratching steadily alongside its male parent for an hour or longer, but never begging for food and never being offered any.

## Display and song

As the young lyrebird becomes more self-reliant, towards the end of the first year, it enters a new phase of its life and begins to sing and display. Initially, displays are brief and sporadic, as if the bird were self-conscious, and the song is very imperfect; but, with practice, both activities improve. Generally, by the third year, displays and song are creditable and many fourth-year birds are proficient in both mimicry and specific lyrebird calls. Immature lyrebirds form loose associations with one another and the two friends often feed together, sing together and even display together on the same mound. The birds often circle one another on the mound, with their tail feathers intermingled and beaks almost touching. They obviously enjoy this experience so much that it continues throughout their lives. Even mature birds in tail moult will join an immature male in a mutual display.

I have not seen two young birds actually fighting, although this does happen; but occasionally have observed an amusing

*A confrontation between Red/Blue and another immature lyrebird.*

confrontation on a mound. The bird Red/Blue was rather bold in some ways — he showed no fear of me and would pluck a piece of cheese from between my teeth. Once he was standing near a mound on which another immature lyrebird was displaying. After several minutes, R/B indicated that he wished to take control of the mound and moved in to displace the other bird. The latter was decidedly not impressed; he lowered his tail, stood his ground and pushed his chest out towards R/B. The intruder paused for a moment, fluffed-up his feathers, raised his crest and reluctantly moved off, as if to say, 'Oh, well, it wasn't all that important; some other time, perhaps.'

On another occasion I was watching R/B (in his fourth year) and Y/Y (Spotty Junior, then in his sixth year), which were 'relating' to one another. R/B was perched on a low stump, engaged in intermittent preening, while Y/Y was singing and displaying on a

mound about five metres away. He was directing his display at R/B, which, however, did not betray much interest. After a while, Y/Y left the mound and stood before R/B with his neck outstretched and singing in a low voice, apparently begging. Suddenly, R/B moved his head in Y/Y's direction and regurgitated the remains of a large worm which Y/Y promptly picked up and swallowed. The two birds then walked away side by side into the forest and were lost to sight.

On 22 December 1964 R/B, now in the process of acquiring his first set of filamentaries, and W/W (another immature male, in his fifth year, but with a plain tail) were together. R/B was singing and displaying on a mound, while W/W stood nearby. From time to time, W/W would run under R/B's outspread tail, but did not remain there. R/B

*The mutual interaction of young lyrebirds is an important part of their development. The older bird (Red/Blue, 5¼ years old) is stimulated by the presence of the younger bird, Double Black, while the latter, in his first tail moult, is encouraged to emulate his friend and soon begins to display. The raised crest signifies nervousness.*

shimmered his tail feathers, just as a mature male does when he is courting a female during the breeding season, but W/W merely kept running back and forth under the tail. When the display ended, after 10–12 minutes, the two birds ran off together and jumped up on a large log. W/W remained stationary while R/B serenaded him from a distance of less than a metre. After several minutes, they descended and ran off into the forest where R/B continued his amorous advances to W/W. Two days later, Red/Blue was visited on the mound by Double Black which had recently shed all his tail feathers, at the beginning of his first tail moult.

Numerous other examples could be given of the social behaviour of the lyrebird, especially immature birds, which often scratch in the same hole while seeking food together, roost alongside one another or in close proximity and, sometimes even share a bathing pool together.

## Changes in throat coloration

During the last two weeks (approximately) of the chick's occupancy of the nest, feathers having a decidedly rufous colour develop on the forehead (frons), sides of the head above the eyes and on the throat. By fledging time, the rufous coloration is usually quite pronounced; but, occasionally, it is only beginning to appear at this time and probably develops fully after fledging. I have not seen a grown chick in its first year without the rufous coloration. The immature birds, like the adults, undergo a 'head-and-throat' moult in January–March each year, when these feathers are renewed over a period of about a month. The lyrebird looks forlorn at the time, but obviously does not feel dejected, because the immature birds and mature males sing and display actively. At the end of the moult, the head and throat are sleek-looking. The frons coloration is not obvious after the second moult, but the throat coloration is at its peak intensity during the second and third years, fading somewhat after the third moult and is usually not evident after the next moult.

Mature birds of both sexes usually lack the throat coloration, but there are exceptions. For example, the bird Blue/White showed a faint coloration in his eighth year, but it was absent after the next moult.

The significance of the rufous throat coloration was apparently not recognized by some early observers, who sometimes confused immature birds with adult females, because of the similarity of the tails. The 'adult females' described by Gould (1848) and Mathews (1919) were without doubt immature birds, but these errors seem to have passed unnoticed.

# 7 | The lyrebird's tail

My studies on the development of the lyrebird's tail began in 1945, but progress was slow because of the difficulty of positively identifying subjects or because I lost track of them as they moved out of my study area or succumbed to predation. It was not until 1958, when the Sherbrooke Survey Group (Watson, 1958) began its banding programme in Sherbrooke Forest, that the first obstacle was overcome. The Group has published several papers on various aspects of the lyrebirds, including plumage changes (Maroney, 1970), nesting (Reilly, 1970), and polygyny (Kenyon, 1972).

It was not possible for me to participate in the organized field work of the Group, but I continued my work as opportunity permitted and the work to be described extended over many years (Smith, 1968, 1982).

Over the years I have collected numerous lyrebird tail feathers resulting from natural moults in the forest; I have also collected almost complete sets of tail feathers from birds in their first, second, third and fourth years, which had been destroyed by predators. I have also examined many skins in the lyrebird collections at the National Museum (Melbourne), the American Museum of Natural History (New York) and the British Museum of Natural History (Tring, Herts.), as well as observing live birds (mainly in Sherbrooke Forest) in various stages of development.

During his visit to Australia in 1838–40, John Gould made a special study of the lyrebird, but his remarks on the moulting of the tail feathers are confusing. In his *Birds of Australia* (1848) Gould wrote,

'all the beauty of this bird lies in the plumage of its tail, the new feathers of which appear in February or March, but do not reach their full beauty and perfection until June; during the next four months, it is in its finest state; after this the feathers are gradually shed, to be resumed at the period stated.'

Unless moulting schedules have changed dramatically during the past 140 years, Gould's observations are at variance with what we know today. Gould's stay was too short for him to make extensive observations.

Another to contribute to the confusion was Tregellas (1930) who stated that 'the birds moult by the first day of June; the old feathers loosen and fall out and the new quills appear. No two birds moult alike'. 'The new ones grow rapidly and by the end of July are in fine fettle.' According to Tregellas's time-table for the tail moult, the lyrebird would be bereft of his tail feathers at the very time when the birds are at the peak of their courtship activities.

The age at which a lyrebird matures has been the subject of some conjecture; in what follows immediately the authors were probably referring to the age at which the male bird acquires his first complete set of filamentary feathers. Williams (1883) considered that the lyrebird took three years to mature; Miss Gillan (1924) and Ambrose Pratt (1933) thought that the bird matured in his fourth year; David Fleay (1942) suggested that the bird matured in his fifth year, while Godfrey (1905) described a semi-captive lyrebird (Jack, which lived on a farm in Gippsland for twenty years from 1885) which 'acquired a magnificent tail at the age of 6–7 years'. Somewhat surprisingly, J.C. Mahan (1906) concluded that the lyrebird matured at the age of eight years, but adduced no evidence to support this view. It will be seen therefore that there was a dearth of information on the subject.

None of the immature male lyrebirds which I studied appeared to have acquired a mate before assuming full male plumage; however, Campbell and Gray (1941) reported that one particular male lyrebird successfully mated with a female, which duly laid a fertile egg and produced a healthy chick, before he had acquired his first set of filamentaries. Similar behaviour has been observed in some other species, e.g., satin bowerbirds (Marshall, 1954), wrens (Cayley, 1949), peacock pheasants (Delacour, 1951). Some hawks have been known to reach sexual maturity before acquiring full adult plumage.

# General structure of the lyrebird's tail

The lyrebird's tail consists of sixteen feathers (rectrices) arranged symmetrically in two groups of eight about the long axis of the body. The two outer ones are the 'lyrates'; the two innermost ones are the 'medians' and the remaining twelve are the 'main rectrices'.

The lyrate feathers of the mature male have twenty to twenty-four V-shaped bars ('windows') along the inner vane, the first two (i.e., at the distal end) being less perfect than the others. The windows arise because, in those parts where the barbs traverse the bars, the barbules are absent, causing the bars to appear transparent. The barbs in the outer vane vary in length from about 6–8 mm (or less) to about 20 mm, while those of the inner vane are often 85–100 mm long. The distal ends of the lyrates are black and club-shaped, and the distal ends of the barbs of the inner vane are dark brownish-black to black. The borders of the windows and the barbs which cross them are yellow-orange-brown-chestnut in colour. The remainder of the underside is silvery-white.

The average values for the lengths, weights and areas of the lyrates of two mature male superb lyrebirds ('A' and 'B') along with the average lengths, weights and widths of black tips of thirteen lyrates from different

mature males are given in Table 1 (p. 82), which includes also data for a lyrate designated 'C', having a length (697 mm) and a weight (2.36 g) near the middle of the range of the respective values for the thirteen mature male lyrate feathers referred to above.

The medians are the longest, narrowest and most curved feathers in the tail. When the tail is folded, as in the normal position, the medians extend beyond the other feathers and their distal ends curve downwards toward the ground and outwards away from the long axis. The average length of thirteen medians was 732 mm (range, 682–792 mm) and the average weight was 0.59 g (0.54–0.67 g).

Because of their filamentous structure, the main rectrices of the mature male lyrebird are known as 'filamentaries'. Over most of the

*This picture shows the detailed structure of the distal end of a mature male lyrebird* (right) *and that of an immature male (White/White) in his sixth year.*

length of the feather, the barbs are separated by spaces of about 5–15 mm; but, proximally to the base, they are contiguous and interlocked through barbules. The extent of this part of the feather is greatest in the middle feathers and least in the outer ones. The feather near the long axis is designated No. 1 and the one near the lyrate is No. 6. The filamentaries of different birds, collected from the forest floor, were separated into groups of 'outer' (Nos 5 and 6), 'in between' (Nos 3 and 4) and 'middle' (Nos 1 and 2). The average lengths and weights for these three groups are given on p. 83.

The barbs of filamentaries vary greatly in length; in a given feather, the barbs near the base may be 85 mm long while those nearer the distal end may reach 175 mm, falling to 15 mm at the tip. In some filamentaries, barbs up to 240 mm were found. The lengths and weights of two almost complete sets ('A' and 'B') of filamentaries are given in Table 2 (p. 84).

# Tail moult

The mature male lyrebird undergoes a complete tail moult every spring (August–December). All sixteen feathers are shed over a period of one to three weeks, usually within ten days. The onset of the moult varies with the bird; some have already shed their tails before the end of August while others may not

## Table 1: Dimensions and weights of lyrate feathers of lyrebirds at different ages

| Age (year) | N | Length (mm) | Area (cm²) | Weight (g) | Width of tip (mm) |
|---|---|---|---|---|---|
| Fourth | 4 | 375 (363–402) | 140 (120–160) | 0.80 (0.73–0.84) | — |
| Fifth | 4 | 410 (402–415) | 160 (150–170) | 0.88 (0.86–0.91) | — |
| Sixth | 3 | 547 (530–563) | 270 (260–280) | 1.33 (1.30–1.35) | 44 (41–45) |
| Seventh | 6 | 600 (585–619) | 330 (310–350) | 1.64 (1.58–1.70) | 52 (46–60) |
| Eighth | 1 | 678 | 370 | 2.01 | 49 |
| Mature | | | | | |
| A | | 640 | 390 | 2.11 | 63 |
| B | | 670 | 380 | 2.33 | 60 |
| C | | 697 | 440 | 2.36 | 70 |
| D | 13 | 690 (638–740) | — | 2.39 (2.09–2.77) | 67 (60–76) |

Values are means, with ranges in parentheses. Of the mature feathers, 'A' and 'B' are from Table 2; 'C' is the feather referred to on page 81; 'D' is the mean value for the 13 feathers referred to on pages 80–81.

do so before November–December. Close observations on one bird (Spotty) in 1954 indicated that the time required to grow a new set of tail feathers was about fourteen weeks. In some immature birds, the time required was about 12–15 weeks, depending on the bird and the number of feathers involved.

The tail feathers of the female superb lyrebird were described on p. 39. Clearly, the two sexes differ markedly in regard to the structure of their tail feathers and there is strong evidence to suggest that the tail moult of the female is protracted, like that of immature birds. Many females have their tails intact while engaged in nesting activities, but I have seen different females, while tending the chick in the nest, to have shed one or both lyrates, one or both medians and several of the main rectrices. Females have been seen growing new lyrates in March and medians in March and April. It is known that these birds were females, because they were accompanied by their chicks of the previous season. The chicks' tails often appear to be longer than those of their mothers, because the tail feathers of the latter are still growing.

One female was seen in complete tail moult in September 1967 while tending a chick in the nest. This could have been a 'shock' moult, precipitated by a confrontation with a predator (fox). The new tail grew normally.

During its first year, the young lyrebird's tail bears a superficial resemblance to that of the female, but there are important differences. The lyrates are barred, but lack windows; also, the first-year lyrates are much smaller than those of the female. The medians and main rectrices of the first-year bird taper more sharply at the distal ends than do those of the female (though I once saw a female whose medians and main rectrices tapered much more sharply than usual). The margins of the main rectrices are often crinkled, a feature which appears to be absent from the female's feathers.

At the end of the first year, the tail feathers are moulted; the medians and main rectrices of the second-year bird closely resemble those of the first year, but the lyrates have a definite form and structure which are diagnostic for the second-year bird. The windows, about sixteen in number, though not as 'clear' as they will be in later years, are more definite and more heavily pigmented than in the first year.

Another tail moult occurs at the beginning of the third year, the new feathers being very different from their predecessors. The lyrates are larger, more heavily pigmented, have clearer windows and have a characteristic shape at the distal end. The medians are longer, more asymmetric than previous medians and taper less sharply at the distal ends. The main rectrices are plain broad feathers having rounded ends, and the degree of asymmetry is greater in the outer feathers than in the middle ones.

The beginning of the fourth year sees another tail moult, the changes which are evident in the third year being accentuated in the fourth. The windows of the lyrates frequently extend right up to the shaft.

## Mature male filamentaries

| Feather group | N | Length (mm) | | Weight (g) | |
|---|---|---|---|---|---|
| | | Mean | Range | Mean | Range |
| Outer | 32 | 658 | 605–705 | 0.78 | 0.68–0.93 |
| In between | 18 | 644 | 600–690 | 0.83 | 0.71–0.90 |
| Middle | 15 | 652 | 625–719 | 0.86 | 0.81–0.92 |

# Table 2: Characteristics of tail feathers of immature and mature male lyrebirds

Values are means, with ranges in parentheses.

| Feather | N | Length (mm) | Area (cm²) | Weight (g) |
|---|---|---|---|---|
| *Lyrates* | | | | |
| First year | 1 | 275 | 80 | 0.35 |
| Second year | 6 | 271 | 80 | 0.37 |
| Third year | 3 | 304 | 110 | 0.52 |
| Fourth year | 1 | 366 | 160 | 0.84 |
| Mature set 'A' | | 640 | 390 | 2.11 |
| Mature set 'B' | | 670 | 380 | 2.33 |
| *Medians* | | | | |
| First year | 2 | 445 | 130 | 0.55 |
| Second year | 2 | 434 | 110 | 0.56 |
| Third year | 1 | 489 | 130 | 0.68 |
| Fourth year | 1 | 526 | 170 | 0.83 |
| 'Adolescent' Type 'A' — 5th–7th year | 6 | 567 | | 0.52 |
| Type 'B' | 2 | 679 | | 0.59 |
| Mature set 'A' | | 682 | | 0.55 |
| Mature set 'B' | | 732 | | 0.59 |
| *Main rectrices* | | | | |
| First year | | 350 (295–400) | 120 (90–150) | 0.48 (0.37–0.60) |
| Second year | | 367 (310–423) | 130 (90–170) | 0.55 (0.42–0.68) |
| Third year | | 366 (310–410) | 150 (110–190) | 0.62 (0.51–0.72) |
| Fourth year | | 393 (350–454) | 170 (140–220) | 0.78 (0.66–0.94) |
| Mature set 'A' | | 620 (606–632) | | 0.79 (0.73–0.85) |
| Mature set 'B' | | 644 (632–660) | | 0.77 (0.73–0.81) |

Data pertaining to the dimensions of the several types of tail feathers of birds in their first to fourth years are given in Table 2. The weights of complete sets of tail feathers of lyrebirds in their first to fourth years are set out below, along with comparative data for adult birds 'A' and 'B'.

| Year | Total weight of 16 feathers (g) |
|---|---|
| First | 7.55 |
| Second | 8.49 |
| Third | 9.78 |
| Fourth | 12.69 |
| Mature 'A' | 14.82 |
| Mature 'B' | 15.11 |

# Later developments

The tail-moulting process is repeated annually; but, because lyrebirds differ so markedly in their moulting behaviour and because each follicle acts independently of the others, according to its own time-table, it is not possible to generalize and some remarkable combinations of the several types of feathers are frequently seen. Not only are feathers of the same type out of phase with respect to time but also, often, with respect to form. Thus, at the conclusion of the moult of a particular bird, the lyrates may be dissimilar, medians may be different and the main rectrices may differ in that some may be plain feathers while others have advanced to the filamentary stage. While it is not possible to describe every one of the many combinations which have been observed, a brief account of the development of the three types of feathers will be given.

The age at which the first filamentary is grown, replacing a plain feather, varies from bird to bird; for example, the birds Bk/Bk/W

An immature lyrebird displaying; the lyrates and medians are second-generation feathers; the first four plain feathers on the left are second-generation feathers, while the remainder, with rounded ends, are third-generation, showing that a tail moult is in progress early in the bird's third year.

*Red/Blue grew his first filamentary feather in his fourth year and his adolescent medians in his fifth year. His lyrates are out of phase structurally; the left lyrate is longer than the other and has a black tip (immature) which is lacking in the right lyrate. Here he is shown displaying, at the age of four years and nine months.*

(hatched August 1967) and W/R/R (hatched August 1971) grew their first filamentary in their second years. I have not seen a filamentary feather in the tail of a first-year bird, but would not exclude the possibility of such an occurrence.

A young male chick found beside a forest road in January 1941 and taken to the Healesville Sanctuary had lost some of its tail feathers and new ones were growing in their places (Fleay, 1941). Among the new feathers were 'several filmy plumes'; thus, as the result of an accidentally induced (partial) tail moult, certain follicles were producing filamentary feathers, although the chick was less than three months old.

In 1866 Dr J.E. Grey, FRS, at a meeting of the Zoological Society of London, read a

*A lyrebird chick aged thirty-nine days, temporarily outside the nest, stretching, showing why the nest needs to be capacious. Note that the wing feathers are still in their sheaths, showing that growth is incomplete, and the bare areas not yet covered by feathers. The tail feathers were about 140 mm long and not yet completely grown.*

*A lyrebird chick aged about two and a half weeks, showing wing feathers emerging from their sheaths.*

*A female lyrebird approaching the nest to feed her chick, with food stored in her cheek pouches and a large worm dangling from her beak.*

*A female lyrebird feeding her chick on the nest. Note that she has used her nictitating membrane to protect her eye from damage by the chick's beak.*

*A female lyrebird, with her chick's dropping in her beak, flies away to a stream 100 metres distant.*

*This chick stood on the platform of the nest for two hours before finally leaving the nest, 2 metres above the ground.*

*A female lyrebird and her six-month-old chick feeding side by side in a forest.*

*A female lyrebird feeding her six-month-old chick.*

letter from Dr George Bennett of Sydney, describing a young lyrebird which had been in the care of Mr J.S. Palmer of New Town, near Sydney, for some two years and which, when seen by Dr Bennett on 4 January 1866, was thought to be 'a fully-grown female or a male in immature plumage'. However, Dr Bennett was able to inform Dr Grey that 'on this day (27th March), he had learned that the bird was developing two of the "peculiar tail feathers" of the male'. The bird was then in his third year.

The birds R/B and R/Y grew their first filamentary feathers in their fourth year; R/B grew one and R/Y grew three. The bird W/W was in his sixth year before his first (seven) filamentaries were grown, while R and B/W were in their respective seventh years before they acquired their first filamentaries. The number of filamentaries grown in the first 'crop' varies with the bird; R/B and Bk/Bk each grew one; Palmer's bird grew two; R/Y grew three; R grew five; an unidentified bird known as 'The Old Changeling' grew six and W/W grew seven.

The first filamentary is not in the ultimate form of this type of feather and one or two further moults will be necessary before that stage is reached.

At each moult, the lyrates become longer and broader, with further elaboration of the windows and distal ends, as they approach the mature form. Darkening of the distal end and the assumption of a club shape in one or both lyrates of some birds, e.g., R/B, occurs as early as the fourth year and has been observed in an unbanded bird in its third year; or in the fifth year, e.g., W/W. In other cases, e.g., R and B/W, the fifth and sixth generations of lyrates more closely resembled the fourth-year lyrates of younger birds.

I was fortunate enough to obtain one of W/W's sixth-year lyrates which he dropped on a mound after a display and, by comparing this with other feathers, it was possible to identify lyrates thought to be representative of several generations of such feathers. Dimensions and weights of these feathers are given in Table 1 (p. 82), which includes data for a fourth-year

*The first four generations of lyrates; the 'windows' develop in successive generations and the distal end changes shape. The asymmetry of the feather increases as the bird matures.*

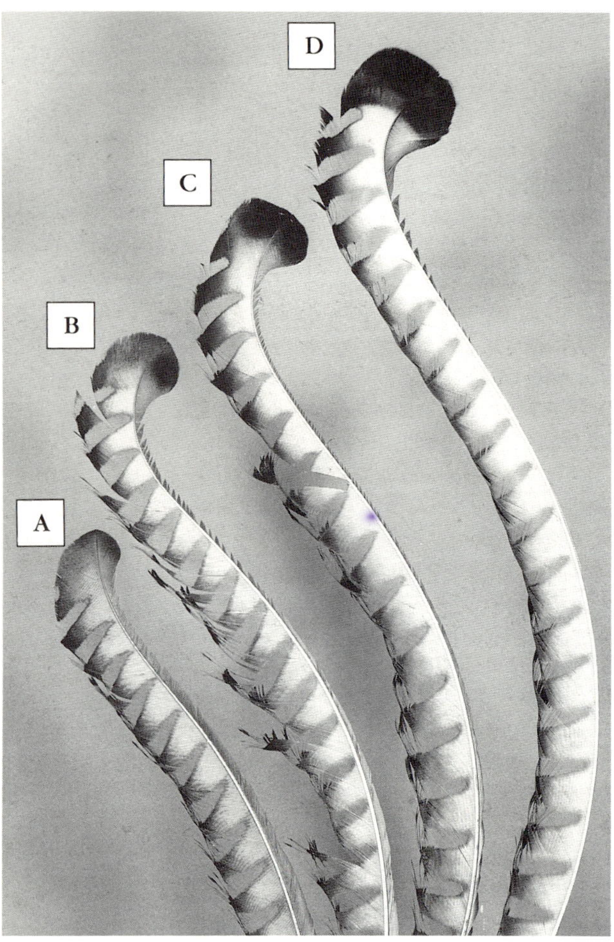

*Later generations of lyrates of the superb lyrebird:* A *fifth generation;* B *sixth generation;* C *seventh generation; and* D *the ultimate form.*

bird and a mature male. It will be seen from Tables 1 and 2 that the lyrates increase steadily in size and weight from being the smallest to the heaviest feathers in the tail.

The first four generations of medians are plain feathers in which there is a steady progression in dimensions, weight and degree of asymmetry at successive moults. Some immature birds, e.g. Y/Y and W/W, produced a fifth generation of plain medians, while R and B/W grew plain medians even in their respective sixth years. However, at some stage in the life of the young male lyrebird, a different type of feather is produced and, because it so clearly marks the transition to the male form, I have called it the 'adolescent' median, for easy reference. The medians of the female superb lyrebird remain similar to those of the fourth-year bird.

The age at which the adolescent median is acquired varies from bird to bird: for example, R/B and R/Y were in their fifth years; W/W and Y/Y were in their sixth years, while R and B/W were in their seventh years before they grew this type of feather. The average length of six of these feathers was 567 mm (range, 523–625 mm) and the average weight was 0.59 g (range, 0.53–0.68 g).

Some male lyrebirds grow a median (Type 'B') which is similar to the adolescent median (Type 'A'), before producing the more usual successor (Type 'C') to the adolescent median. The vane of this intermediate type of median is narrower at both the distal and proximal ends, on both sides of the shaft, than it is in the adolescent median. I have seen this type of feather in two live birds in Sherbrooke Forest and in a specimen at the British Museum of Natural History. In the latter case, it was coexisting with an adolescent median, suggesting that the bird was in the process of moulting when taken. The average length of the two B-type medians collected in the forest was 679 mm (675, 683 mm), and the average weight was 0.56 g (0.53, 0.59 g).

The type 'C' median is similar to type 'B', except that the barbs on the outer side of the distal end are separate. In the ultimate form (type 'D') of the median, the barbs on both sides of the shaft at the distal end are separate, those on the inner side being the longer. Except at the distal end, the barbs on the outer side are extremely short (1–2 mm) and contiguous, while, on the inner side, the barbs are separate and about 5 mm long.

Once the bird has reached the stage of producing the adolescent median, even though this may have been delayed until the seventh year, the three types of tail feathers advance, through a further one or two moults, to the stage at which a complete set of filamentary feathers has been acquired, along with lyrates and medians of structure appropriate for birds of that age. However, in three cases which I have observed, one of the follicles had a plain feather along with eleven filamentaries. At the following moult, a complete set of filamentary feathers was produced; in such cases, some of the follicles would have grown three generations of filamentaries, and others two, before the remaining follicle had produced its first filamentary.

Normally, at each tail moult, all sixteen feathers are renewed, although the moult may have extended over six to seven months. Once growth of a feather has ceased, the follicle remains dormant until the onset of the next moult of that particular feather. The normal period of dormancy of a follicle is therefore about 38–39 weeks; but, in some cases, it may be a year or so longer. The bird 'R', at the end of his sixth year, had a tail consisting of twelve plain feathers (sixth-generation), two lyrates the tips of which lacked black pigment, and two plain (sixth-generation) medians. At the next moult, lyrates were grown which had the beginnings of black tips, along with a pair of adolescent medians and five filamentaries, but the follicles of seven plain feathers remained dormant. These sixth-generation plain feathers were carried through to the next moult, a year later. Even so, these feathers were not the first to be shed, so that, at one stage, Red's tail contained a mixture of sixth-generation plain feathers, some first-generation filamentaries (replacing some of the sixth-generation plain feathers which had been shed), and some second-generation filamentaries (replacing some

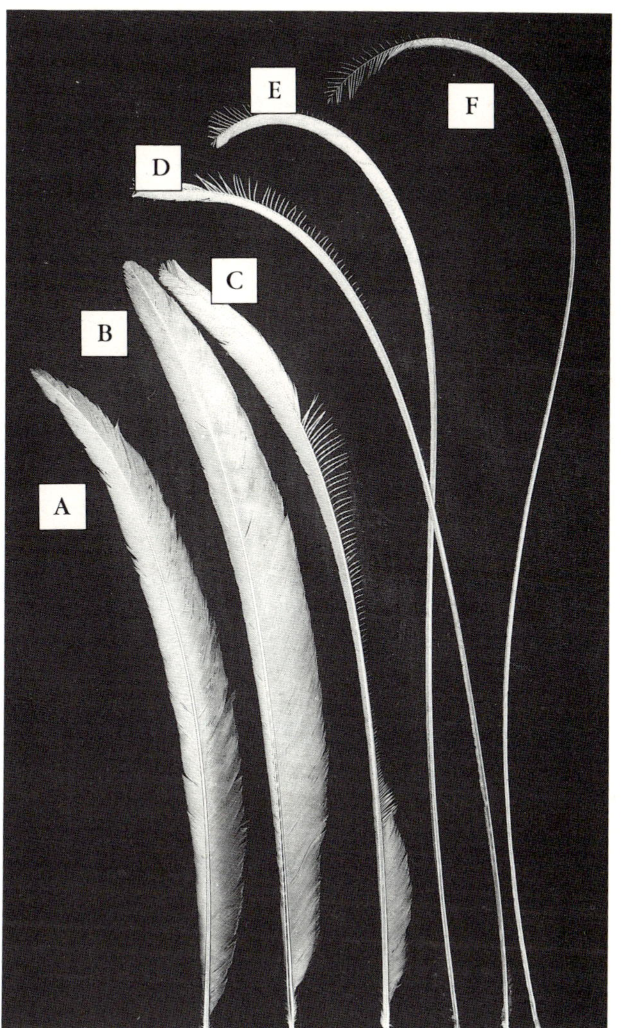

The picture demonstrates how the medians change in size and structure as the bird grows older: A *a first-generation median (second generation is similar)*; B *a third-generation median showing the blunted distal end characteristic of this and later generations of plain medians*; C *the adolescent median*; D *the 'type B' median which in some birds succeeds the adolescent median*; E *the 'type C' median which usually follows the adolescent median*; and F *the ultimate form of the mature male median.*

*Single Red, at this early stage of his sixth tail moult (22 September 1964), had eleven plain feathers, two lyrates lacking pigmentation in the tips, and had shed one (plain) median. The moult was protracted.*

first-generation filamentaries which had been shed) along with the lyrates and medians. At the conclusion of the moult, all sixteen feathers had been renewed and Red had his first complete set of filamentaries, along with two Type–C medians and two Type–C lyrates.

In some lyrebirds, the plain feathers adjacent to the lyrates are distinguished by pigmented bars and, in some cases, windows similar to those of the lyrates. The effect usually extends into opposite pairs of feathers, the intensity of pigmentation diminishing towards the middle of the tail. The degree of barring and pigmentation varies with the bird; in some, the effect is often pronounced even in the third year; in others, it is not apparent until much later.

There are two other types of barring in feathers which have been described by Glegg (1944) and Wood (1950). The more obvious type is the well-known 'fault bars', said to

| Bird colour-banded | R/B | R/Y | R | B/W | W/W |
|---|---|---|---|---|---|
| First filamentary grown in year | 4th | 4th | 7th | 7th | 6th |
| Filamentaries in first 'crop' | 1 | 3 | 7 | 2 | 5 |
| Age when first complete set of filamentaries acquired, year | 6th | 6th | 8th | 9th | 8th |
| Adolescent median grown in year | 5th | 5th | 7th | 7th | 6th |

be developed at night, between the hours of 1 a.m. and 5 a.m., and which are attributed to the lowering of the blood pressure which occurs at that time, because of the consequent reduction in nutriment supplied to the growth centre of the feather. This type of fault is common in the plain feathers of the lyrebird, but I have occasionally observed it in the lyrates and medians of immature birds.

The other type of barring in feathers has been described by Wood (1950) as 'those elusive alternate light and dark normal engular markings across the web of many feathers, having the appearance of watered silk without any sheen'. They are seen only by reflected light and best by tilting the feather. I have observed these growth bars in the plain feathers of immature and female lyrebirds and in the lyrates of adult birds.

The annual tail moult is repeated until a complete set of filamentary feathers along with two lyrates and two medians is produced, except in comparatively rare cases, when one plain feather remains. However, the moult which produces this set of feathers will have been protracted, and the first 'mature-type' moult will occur in the following spring. At this moult, all sixteen feathers will be shed over a short period, usually within ten days, and the new feathers will be in phase with one another.

The differences between birds in regard to some of the principal aspects of tail development are summarized in the table above.

*At the conclusion of his sixth tail moult, Single Red had acquired five filamentaries but retained seven plain feathers from the fifth tail moult. His lyrates had acquired pigmentation (though not yet mature) and the medians were of the adolescent type.*

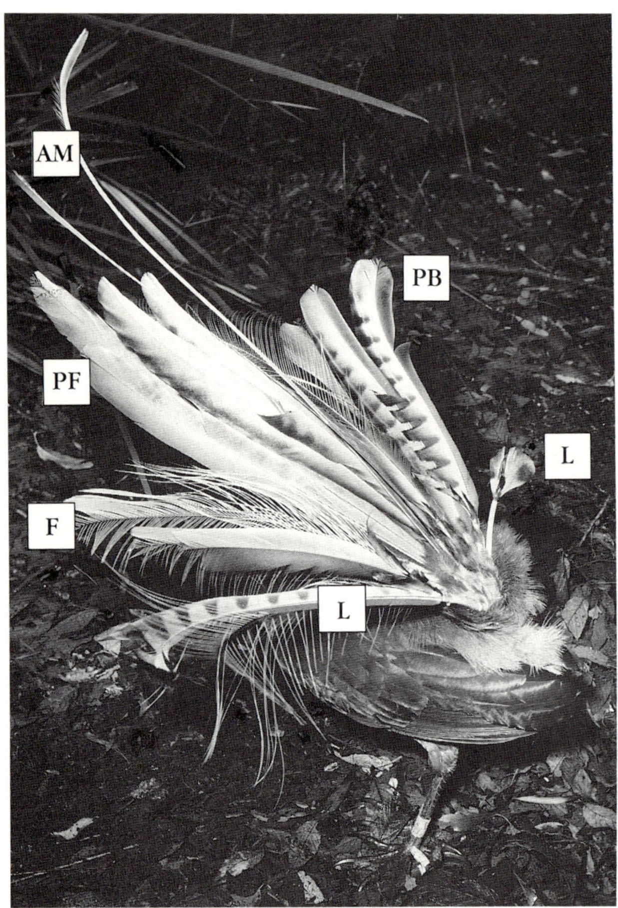

*White/White, on 16 January 1966, growing a new tail:* F = *filamentary;* L = *lyrate;* AM = *adolescent median;* PF = *plain feather;* PB = *barred plain feather.*

# The role of the endocrine system

Although no experimental work has been done on the role of the endocrine system in the development of the tail feathers of the lyrebird, enough information is available on the subject in regard to other bird species to justify the assumption that the same general principles would apply to lyrebirds as have been demonstrated for other species (Voitkevich, 1966; Lillie and Juhn, 1931; Höhn, 1960, etc.).

For each of the feathers comprising the several generations of tail feathers which are produced as the bird matures, there is a 'coded genetic blue-print'. The translation of the information encoded in the cells into the morphological units which we recognize as feathers is complex (Brush and Wyld, 1982) and involves the interaction of many chemicals, including amino-acids, enzymes, minerals and hormones. The principal hormones directly involved in feather production are those produced by the thyroid gland and the gonads, but these bodies depend for their proper functioning on the efficient performance of the other members of the endocrine 'team', including the hypothalamus and the pituitary (Lewin, 1972).

Impairment of the normal functioning of the thyroid gland or the sex organs (experimentally or otherwise) is reflected in imperfections and variations from normal in the shape and size of the feathers being produced at that time. Further, normal sex-gland function demands a normal level of thyroid hormone in the blood (Höhn, 1950). Restoration of the thyroid or gonad function results in the production of feathers in their usual stereotyped forms.

The thyroid hormones (mainly thyroxine) are concerned principally with oxidative processes in cells, i.e., metabolism. They have a special role in the regulation of protein required for feather production, and the steady increase in size of successive generations of tail feathers probably reflects increasing levels of activity of the thyroid gland as the bird grows older.

The sex hormones are principally concerned in reproduction activities, but also have a role in determining feather structure (Lillie and Juhn, 1932; Voitkevich, 1966). A study of the tail feathers of the lyrebird, as presented in the foregoing, leads to the conclusion that the fully vaned (plain) feathers of immature lyrebirds and females are formed through the mediation of the female hormone, as are also those parts of the male feathers in which the barbs are contiguous and interlocked through barbules. The fine barbs of the filamentary feathers and the distal ends of the medians, along with the barbs devoid of barbules in the windows of the lyrate feathers, reflect the mediation of the male hormone during the production of such feathers. The variability in the appearance of male feathers in the lyrebird's tail is no doubt related to varying levels of male hormone in the blood. At maturity, a stable plateau of endocrinal activity is achieved, making it possible, at the appropriate season, to produce the several types of male feathers in their stereotyped forms.

It seems reasonable to assume that the onset of the annual tail moult is initiated by the hypothalamus–pituitary system which is activated by the CNS which responds to the triggering of the pineal gland by environmental factors. The moulting process results from the pushing out of the old feather by the developing new one, so that the moult in birds is a single process actively concerned only with the production of feathers of the new generation (Watson, 1963).

The thyroid gland plays an important part in the pigmentation of feathers (Voitkevich, 1966). By injecting with thyroxine, and thereby increasing the level of the blood, Lillie and Juhn (1932) demonstrated the relationship between hyperthyroidism and the formation of black melanin (eumelanin) in the saddle feathers of the brown leghorn fowl. It has been demonstrated that the jet black pigment in the English blackbird, in the fur of black rabbits and in human hair is black melanin (Fox and Vevers, 1960). However, this melanization proceeds only in the presence of tyrosine and the enzyme tyrosinase. The thyroid hormones thyroxine and tri-iodo-thyronine are closely related to the amino-acid tyrosine. It seems very probable that the jet black pigment in the tips of male and near-mature male lyrebirds is due to eumelanin, formed in consequence of the requisite degree of hyperthyroidism, at the time these parts of the feather were being formed. The yellow-orange-brown-chestnut pigment observed in the lyrate feathers is phae-melanin and indicates that the concentration of thyroid hormone in the blood when that part of the feather was being formed was lower than it was when the black tips were being formed.

The pigmentation of the main rectrices adjacent to the lyrates is therefore probably due to hyperthyroidism, but the colour is yellow-orange-brown-chestnut, due to phae-melanin, as distinct from the eumelanin (black) of the margins of the barbs of the lyrates of mature males and the black tips of the lyrates.

**8** | # The song of the lyrebird

In a broad sense, the songs of birds, including that of the lyrebird, take the place of speech in humans, providing a means of communication between birds of the same and of different species.

The lyrebird is only one of about fifty different species of birds listed by Chisholm (1965) as being noted for their mimicry, but the lyrebird alone is renowned as the 'Prince of Mocking Birds' and is said to be superior in mimicry to the mocking bird of North America.

The lyrebird, like all great artists, achieves perfection only through the assiduous cultivation of a real (inherited) flair for producing sounds having musical quality, and the lyrebird's vocabulary develops as the bird grows from a chick to maturity.

## The young lyrebird

Until the lyrebird chick's eyes are open, it is largely silent. Once its eyes are open, it is able to distinguish between its mother (which has biological rights to the nest) and an intruder. While being fed, the chick emits soft wheezings and gulps reminiscent of the gurglings of a contented human baby. However, if an intruder (such as the author) places his hand near the entrance of the nest, the chick emits a spine-chilling shrill screech calculated to set the inexperienced intruder back on his heels; but (this is a trade secret), when one has recovered from the initial shock, if one places the hand over the chick's head and strokes it gently, it settles down in the nest and makes a soft cheeping sound which obviously expresses contentment similar to the purring of a cat. Presumably, the hand-stroking reminds the chick of the gentle pressure of its mother's body and evokes a pleasant sensation.

During the chick's occupancy of the nest, there develops a pattern of conversation between it and its mother; the latter approaches the silent nest with a series of soft 'oo...oo...oo...' notes (as in 'look') and the chick responds with its soft contented cheepings. As the chick grows older, say, from two and a half weeks, the volume of the alarm call increases; but, of course, this is uttered only in self-defence. It is a different alarm call from that used by older birds, presumably to enable the parent bird to recognize the chick's distress signal, should this be necessary. The growing chick may make a note which sounds like 'tchick...tchick' or the sound humans produce by exhaling though the open mouth (try it), if an intruder disturbs it by placing the hand in the nest. This is resisted with vigorous thrusts of the chick's large claws; but, if the intruder persists, he very quickly receives the full alarm call. In addition, the chick may utter a harsh guttural sound which discourages all but the brave.

The chick emits a shrill piping note to inform its mother that it needs assistance with its toilet. The limited vocabulary of the nestling is designed to ensure that it receives the necessary attention from its sole supporting parent and to protect it against predators. That is all it needs at this stage of development.

These are the principal items of the vocabulary of the fledgling, but expressed more vigorously. All these calls, in a general way, are related to survival behaviour.

During the first few days after leaving the nest, the young lyrebird is largely silent. This ensures that its hiding place is not advertised to predators. After it has acquired the habit of using perches, especially when it has reached the stage of 'climbing' up into the crown of an olearia tree, it makes a piping note like 'eeeeeeek' which, however, differs from the shrill alarm call which accompanies the chick's flight to safety if it is disturbed.

When the young bird has reached the stage of accompanying its mother on her never-ending foragings, the two birds 'talk' to one another almost non-stop. The young bird makes a soft wheezing sound, while the female makes the 'oo...oo...oo...oo...' note.

While continuing the soft conversational wheezings to maintain contact with its mother, and the shrill alarm call to signal an emergency, the young lyrebird begins to develop its prowess as a mimic at an early age. I do not know just when the chick begins to give expression to the notes it has been storing in its memory bank since leaving the egg, and the earliest record I have is of an eight-month-old chick which gave a very good imitation of the staccato barking of a fox-terrier, while in the company of its two parents, one morning in April 1939.

As Professor Thorpe (1961) explains, learning bird-song is largely a mimetic process in which the bird, by a trial and error approach, endeavours to reproduce sounds through the actuation of the vocal motor-mechanism with which it is naturally endowed. There is an interesting story behind the chick's imitation of the fox-terrier; the dog had been taken into the area near the Falls in Sherbrooke Forest by a forest worker, and the lyrebirds nearby had quickly learned to imitate the barking of the terrier. On the morning of my visit, five years after the departure of the dog, the young bird's male parent was imitating the dog's barking and the young bird was imitating its parent. To judge from the quality of the imitation, it seemed that this was not the first time. The chick obviously enjoyed making this note, because it repeated it often — in fact, it was the only note it uttered, apart from its regular conversation with its mother.

It is difficult to describe the sounds made by young lyrebirds in human language and, in attempting to do so, one runs the risk of being accused of indulging in what the late Professor Jock Marshall called 'anthropomorphic nonsense'. However, one sometimes sees and hears a seven- to eight-month-old lyrebird, perched on a low branch while its parents are feeding below, making a series of loud staccato squawks in rapid succession, which remind me of the noises made by Donald Duck when he is voicing his frustrations over the misdemeanours of his celebrated nephews. There is no 'melody' in these utterances, which are highly repetitious, but occasionally one hears an imitation of the barking of a dog. The bird is not alarmed, or it would flee in haste; these noises may simply be 'lyrebird baby talk'; but they are certainly amusing to human ears.

When Spotty's 1957 chick was about thirteen months old, he was seen making very tentative displays on some of his father's mounds and singing in a very juvenile manner. It was difficult to identify any particular notes at the time; but, a few years later, he was on the way to emulating his illustrious father in both mimicry and specific lyrebird calls.

Once the young lyrebird begins to sing and display, impelled no doubt by seeing and hearing its companions doing likewise, improvement comes rapidly and, by its third year, a young lyrebird is singing strongly. Through continual practice, the bird improves the quality of its song and repertoire which now includes most of the notes of the mature bird, except those used in courtship. Even the territorial call is produced, although it has no biological significance as yet, except that, because of its inclusion in the bird's repertoire at this early age, it will be produced more proficiently when it is required later.

Not all birds are equally skilful in mastering the notes or calls which are later to become their stock in trade. Some birds (e.g. Red) have great difficulty with the notes of the whipbird and the pied currawong, and mimicking the chuckling of the kookaburra requires much practice; but eventually the notes flow smoothly from the bird's throat.

Immature birds frequently engage in group singing, especially early in the morning. One of my best recordings is of a trio of immature birds which were singing together one morning in the autumn of 1965. Each bird was doing its own act, but often picked up the call of one of the other birds, especially when one of them was imitating the whipbird or grey thrush. Their timing was perfect, as if they had but one voice. Immature birds often display together on the same mound, virtually singing down one anothers' throats. Sometimes, one bird will occupy the mound and sing and display to the other, standing in front of him.

# Vocal behaviour of the female lyrebird

Normally, the female lyrebird is not very vocal, but she does have a number of different calls which she uses in particular situations. If disturbed in the forest, she emits a shrill alarm call as she flees to safety. If she is in doubt, but not actually alarmed, she is likely to make a sound which (to me) is like 'aw-kok', the second syllable being in a higher key than the first. When in attendance at the nest and later, in the forest, she communicates with the chick by means of a soft 'oo…oo…oo…' note.

However, while the adult female does not normally sing and display as immature birds and mature males do, when she is under stress, she is much more vocal. For example, in October 1938, while I was inspecting a chick in its ground-level nest, I inadvertently caused it to fly out of the nest, to the accompaniment of loud squawks and shrill alarm calls. The chick came to rest in a shallow pool in the creek, where it sat with its wings spread on the surface of the water. While I scrambled down the bank, heaping coals of fire on my head for having been so careless, the female was bounding about and perching on the trunks of tree-ferns, squawking and uttering the alarm call. The general cacophony certainly added to my feelings of guilt. Suddenly, the chick became silent and the female, from the crown of a tree-fern, began to sing beautifully. Her mimicry was very clear and she had an extensive repertoire, similar to that of the male. I retrieved the chick, dried it thoroughly and returned it to the nest, where it sat quietly. Within a few minutes, peace and quiet had been restored and the female began to scratch for food as usual, but keeping a sharp eye on me, nevertheless.

Next morning, at 7.30, I found the nest empty. After searching fruitlessly for an hour or more, I heard a solitary 'aw-kok' call from the slope above, about 100 m away and, within a few minutes, had located the female on a large mountain ash which had fallen many years ago. She ran up and down the trunk, squawking and making the alarm call, and endeavoured to entice me away from the log. Eventually, I found the chick hidden under the log on which the lyrebird and I had been trying to outwit one another.

In 1939 I made the acquaintance of the 'Singing Hen', which was so named because, when one ventured too close to her nest, she displayed and sang, just as immature lyrebirds do. Sometimes she performed on the ground, running around with her tail inverted over her head, tripping over obstacles, and sometimes she sang and displayed from the branch of a tree near the nest.

Other examples could be given, and it would probably be fair to say that, under stress conditions, all female lyrebirds would sing and display; but, with experience, I learned to avoid putting them under acute stress, so that, while I was often confronted with a threat display, not every bird actually sang after the manner of the 'Singing Hen'.

The singing of the female under stress is so good that it seems unlikely that she could have achieved such perfection in her first stress situation and it is probable, I think, that she acquired these imitatory skills while she was immature. It is difficult to distinguish between females and immature males in the forest, unless one recognizes a specific male feature such as a filamentary feather or male-type lyrate or median feather in the tail.

This raises the question of why the female, having become as proficient at singing and displaying as an immature male, should have abandoned these talents after having matured, except under stress. The answer is probably that to have continued to practise these forms of behaviour would have proved disadvantageous to the species and that therefore, through the process of maturation, the endocrinal system developed in her a pattern of behaviour which was advantageous to the species; that is, the ratio of female to male sex hormone in the blood of the normal female was favourable to the accentuation of female behaviour and the suppression of secondary male characteristics. However, in some females, the ratio of male to female sex hormone in the blood may be abnormally high, resulting in the acquisition of secondary male characteristics, including singing and, occasionally, several male-type (i.e., filamentary) feathers in the tail.

It would be confusing if lyrebirds other than breeding males were to engage in loud vocalizations and displays during the breeding season, which no doubt explains also why immature males (with few exceptions) do not sing and display at this time, but are active during the non-breeding season.

# The mature male lyrebird

By the time the male lyrebird has matured, he is an accomplished songster. His extensive repertoire contains the following items:

1. The alarm call — 'eeeeeeeek' or 'whisk whisk'.
2. The 'aw-kok' call (as for the female).
3. A soft noise, like the whimpering of a dog, which is used to express concern or inquiry, but not alarm. I have heard this when a mature male has received a message from a remote source; the bird seemed uncertain

as to what to do next and stood looking intently in the direction from which the message had come, but eventually resumed his preening.

4. Several specific calls, viz:

(a) the territorial call; this is a series of loud notes which are subject to some remarkable regional variations which transcend the dialectal differences between those of some of the species mimicked, in different parts of the range of the lyrebird, e.g., the whip-bird or grey butcherbird from Sherbrooke and Gippsland or south-eastern Queensland. The essential characteristics of these calls remain clearly recognizable, but the territorial calls of lyrebirds from different districts often appear to have little in common.

(b) The 'scissors...scissors...scissors...' item.

(c) The 'blick...blick...blick...' item, which is usually the prelude to a display or coition.

(d) The 'clonk clonk' item, which may be given alone or as part of a more extended call, like 'clonk clonk clickety clickety click', with dialectal variations, which usually accompanies a series of jumps from one foot to the other.

(e) A 'clack...clack...clack...' note which is given at the end of a display when the bird remains on the mound with his tail erect and vibrating. The display is often resumed, sometimes on another mound, to the accompaniment of full song.

(f) Courtship calls, including the 'whisper song', chortling and a remarkable sequence of notes which sound like 'eeaweeaweeaw...'. Of course, it is difficult to find a phonetic equivalent of every item in the lyrebird's song.

(g) Sometimes, as he stands on the mound, between bursts of song, he utters a note like 'parrrrt' or 'warrrrh', which may be repeated several times.

(h) A sound like 'ugh...ugh...ugh...' (not to be confused with the 'oo...oo...' of the conversational female), which is

made after the pre-mating song has ceased and may arise from the effort involved in the act of coition. I have heard and recorded this item several times in the early part of the breeding season, after a female has joined the male on the mound, but I have not actually witnessed the incident.

5. Imitated calls, which are described below.

# Imitated calls

It has been said that the lyrebird can imitate any call it hears, including 'every note of the bush' (Rawnsley, 1863) and a number of mechanical sounds such as the blow of the axe on a tree, the sharpening of a saw, the clinking of chains, the 'toot' or 'choo...choo...' of a steam train, and other noises. There is an amusing story (Bill, 1933) about a lyrebird in Gippsland which learned to imitate the whistle of a chainman engaged in surveying in the district, and which caused much confusion by his ill-timed mimicry. While I have not heard all of these imitations, I have heard a sound which resembled the rattling of chains, in the mimicry of some Tasmanian lyrebirds, but cannot identify the model. However, I have heard the lyrebird imitating the barking of a fox terrier, and once heard a lyrebird in Sherbrooke giving a perfect imitation of the panting of a large dog which had obviously been trespassing in the forest. Arthur Groom, co-founder of the celebrated Binna Burra guest-house on the fringe of Lamington National Park, in the *Sydney Morning Herald* (20 July 1929) tells how he was temporarily hoaxed by the perfect fidelity of an Albert lyrebird's imitation of the barking of a dog, deep in the Albert Gorge in south-eastern Queensland. Tregellas remarked that the lyrebirds in the vicinity of his log cabin in Sherbrooke Forest imitated the yelp of the fox, and Dr Brooke Nicholls (1911) included in the repertoire of a lyrebird from the Bass Valley in Gippsland such items as the notes of the whistling eagle, the boobook owl and even the grunting of the koala.

Most of the imitated calls which I have heard in the songs of lyrebirds have been bird calls, and these varied with the district. For example, the lyrebirds of Sherbrooke never imitate the satin bowerbird, because that bird does not occur in Sherbrooke; but, in certain parts of Gippsland, New South Wales and Queensland, where the bowerbird is present, the notes of the bowerbird are featured in the lyrebird's mimicry. The lyrebirds of Mount Buffalo include several notes of the white-eared honeyeater; while, in other districts, this bird is imitated infrequently or not at all.

The general rule is that bird calls which are heard regularly and frequently by the lyrebird will find a place in its mimicry. Thus, the loud calls of the black cockatoo, grey thrush, pilot bird, currawong, kooka-burra and whipbird are frequently heard in the lyrebird's mimicry. However, while some of these calls, such as those of the whipbird, pilot bird, grey thrush, crimson rosella, kookaburra, black cockatoo, pied currawong, satin bowerbird, golden whistler and a few others, are readily recognized in the lyrebird's song, many other imitatory notes, such as those of the yellow robin, silver-eye, grey fan-tail, eastern shrike tit, brown-headed honey-eater and olive whistler, are so fragmentary that only an experienced and sensitive ear is likely to identify them in the rapid outpour-ings of mimicry that flow from the lyrebird's throat. Many of these notes are used so in-frequently that one wonders why the lyrebird bothers to include them at all.

The answer may be that the lyrebird, having heard them from the model at some time, stores them in its memory bank and that they slip out occasionally, for no particular reason. Or, it may be that it is not particularly attracted to such notes because they lack vol-ume and are therefore less useful. This is not, however, to suggest that the lyrebird's song is purely utilitarian. The alarm call and the territorial call of the male, and the conver-sational notes of the female and chick at vari-ous stages of development, are clearly in this category, but many of the mimetic com-ponents of the song of the male appear (to the

*A mature male lyrebird singing, imitating the black cockatoo.*

human ear) to transcend the purely utilitarian requirements of the courtship ritual.

Often, when a lyrebird is singing, it will intercept a call from a model such as a currawong, whipbird, pilot bird, grey thrush or crimson rosella, and it will reply to the model (which may be some distance away), even in the middle of another item of mimicry, and then pick up the unfinished note as if the interruption had not occurred.

Some of the more unusual items in the lyrebird's mimicry include the rustling of the wings of a flock of crimson rosellas, the flight noises of the yellow robin and grey thrush, and the snapping of the beak of the yellow robin, wattlebird and kookaburra. It is rather amusing also to hear a lyrebird imitating the alarm call of its own species and the noise made by a lyrebird chick while it is being fed by its mother.

The lyrebird's song is loud and penetrating, and highly directional. This ensures that the voice will carry through the dense forest where, even over a short distance, it is usually impossible for visual communication to occur between the songster and the subject for which the song is intended. The percussion notes from such calls as those of the whipbird, golden whistler, pilot bird, etc., often produce a tingling sensation in the human ear and cause 'blowing' in a tape-recorder adjusted for normal reception of other parts of the song.

Many lyrebirds appear to have a decided preference for particular notes; thus, for example, in the Childers district in West Gippsland, the call of the whipbird is frequently imitated with consummate skill and sometimes with delightful improvisations, while the notes of the grey thrush, with variants, comprise a dominant item in the song of Prince Edward's lyrebird in south-eastern Queensland. In other words, some lyrebirds appear to like producing certain notes at a level of musical quality (as judged by human ears) far in excess of basic requirements for courtship purposes.

This preference for particular notes may explain why, fifty years after they were introduced into Tasmania, the lyrebirds there continue to imitate the whipbird, although the latter is not found in Tasmania. In the early 1930s a boy in the Ebor district, in north-eastern New South Wales, had a pet lyrebird which learned to imitate the notes of the flute. When the family left the district, the lyrebird was released to the nearby forest where the other lyrebirds soon learned to imitate the flute notes of the newcomer. This remarkable item of mimicry has persisted in the songs of successive generations of lyrebirds of the district, in New England National Park, and may still be heard.

# Analysing the lyrebird's mimicry

When the lyrebird's song is analysed, it is found that about 70 per cent of it consists of the calls of other birds and animals, augmented perhaps by a few mechanical noises (Robinson 1975; Smith, unpublished data).

The loud ringing notes comprising the territorial component of the lyrebird's song represent only about 11 per cent of the total song, while the 'scissors grinding' and 'clonk clonk clickety clickety click' components together represent about 16 per cent. These figures are subject to wide variation, depending, *inter alia*, on the bird, the time of the year and the district. The territorial item might serve to warn other males and to direct potential mates to the singing male; but I am inclined to think that the female might be more interested in the rhythmical 'clonk clonk clickety clickety click' component. Whether the raucous screeching of the black cockatoo or the rather raspy 'complaining' note of the gang-gang cockatoo would excite the female lyrebird seems doubtful, but it is possible that she responds more warmly to the musical notes of the pilot bird, or the bell-like notes and flock chatter of the crimson rosella, or the melodious notes of the grey thrush, or the polished rendition of the whipbird's call, or the rather cheery notes of the laughing kookaburra. It is difficult to understand just how the inclusion of fragments of the notes of such birds as the yellow robin, grey fantail, silver-eye, magpie or raven could do much to excite a female lyrebird, but perhaps it is the total effect and the fact that, in a prolonged singing bout, the various items (mimicry and specific calls) are not rendered in any set pattern which holds the female's interest.

Of course, it may well be that beauty of song resides in the ear of the listener and it is entirely possible, I think, that the female does not bother to analyse her mate's song at all, and that it merely serves, at the critical moment, to direct her to the mound where their endocrine systems slide into top gear and carry the birds forward on their biological mission.

Among the first to attempt to analyse the song of the superb lyrebird was K.C. Halafoff (1959, 1961, 1962, 1964), who devoted many years to the study of lyrebirds in Sherbrooke Forest and certain parts of Gippsland. Halafoff was well endowed to undertake such work, being both a musician and an engineer. He made tape recordings of the lyrebird's song and conducted elaborate studies on their musical quality. He had no doubt about the 'ability of the lyrebird to give the most perfect imitation of any sound' and that 'it was not a utilitarian motive alone which compelled the lyrebird to arrange a medley of heterogeneous imitations into an harmonious whole, composed similarly to human music' (1961*b*).

Thus, the young lyrebird, genetically endowed with musical potential and being subjected from an early age to a wide variety of bird-song, through constant practice, soon acquires a vocabulary which is gradually moulded into a song of exquisite beauty. The lyrebird's song is not rigid in form; the various items of specific song and mimicry are seldom presented in the same order, because the song is composed as the bird goes along and the notes flow effortlessly, the one often leading smoothly into the following one.

In humans, the quality of speech and song is governed by the manner in which we vary the tension on our vocal chords and by the proper use of the mouth, lips, tongue, nasal passages and breathing apparatus. Proficiency in the production of good speech does not come without some practice and a knowledge of the principles involved. Likewise, it is the controlled use of the mandibles (the upper one is fixed, of course), throat, nasal cavities and other organs which enables the lyrebird to produce whatever note he has in mind at that instant. This, of course, raises the question of whether his articulations are directed by the brain or whether they are produced automatically.

For a variety of reasons, consideration of this aspect is beyond the scope of this book, but I think it is begging the question merely to say that the bird learns to sing or to master a particular sequence of notes by 'trial and error'. Lyrebirds have compelling reasons for persevering in mastering the processes involved in producing the sounds on which the survival of their species depends and, in their forest environment, they have the benefit of tuition from experts. But, even when they are mature birds, they sometimes make mistakes, possibly through a momentary lapse in concentration. Thus, while a bird may faithfully imitate the grey thrush, the whipbird, the crimson rosella and other birds, occasionally, he seems to hesitate and produces a succession of almost unrecognizable fragments of the songs of other birds until, ultimately, his voice flows smoothly from the last note into a perfect imitation of some other bird's call.

I recall how persistent 'Single Red' was when he was learning to imitate the notes of the whipbird; I used to wonder why he persevered. Unless he was somehow conscious of the difference between 'good' notes and 'bad' ones, why should he have continued practising? Somehow, it seemed, the impact of a 'bad' note on his nervous system must have been translated by his brain into a new approach — a change in the manner in which he manipulated his vocal apparatus. Equally, when he received a 'good' signal, he registered that and related it to the particular 'mode' of his vocal apparatus which was responsible for it. Eventually, by this trial and error method, he achieved the desired perfection, but how did he know, when he had reached that plateau, that he was 'there'? While he was practising, he was usually alone; there was no model close at hand, although at some stage he would have the opportunity of comparing the notes he produced with those of the model, like a singing student comparing his own notes with those of a tuning fork.

This, I think, explains why the Tasmanian lyrebirds (see chapter 10) no longer imitate the whipbird with the same fidelity as their mainland cousins; that is, because they are unable to compare their mimicry with the notes of the model.

Lyrebirds generally, having mastered a particular item of mimicry, have the ability to embellish or extend such songs. This is especially noticeable in such items as the grey thrush and whipbird. The lyrebird not only reproduces the notes of the model, but introduces his own 'improvements' or improvisations. This suggests that the lyrebird has the ability to create his own music, as distinct from specific lyrebird calls which are directly related to territorial proclamations and other calls traditional to the species, such as alarm calls, 'come and play' signals, etc.

Hartshorne (1956) has suggested that the lyrebird varies the presentation of his mimicry in order to relieve the monotony of his song, for the benefit of the female. How would the songster know whether the female was bored or agreeably responsive unless he was able to interpret the signals which she presumably gives him?

The lack of a stereotyped pattern in the lyrebird's song makes it difficult, if not impossible, to select a typical passage. Below is a list of the notes produced by one of Sherbrooke's mature male lyrebirds, in sequence, during a three-minute segment of a 30-minute 'concert' recorded in the Broadcast Area of Sherbrooke Forest in June 1967. Space precludes a more detailed presentation, but this may convey some idea of the complexity and variability of the lyrebird's song.

Grey thrush
Yellow robin
Magpie
Territorial
Yellow robin
Grey thrush
Crimson rosella
Black cockatoo
Pilotbird
Wattlebird
Grey thrush
Golden whistler
Yellow robin
Territorial
Whipbird
Wattlebird
Golden whistler
Goshawk(?)
Grey butcherbird
Territorial
English blackbird
Crimson rosella (bell note)
Crimson rosella (flock note)
Whipbird
Magpie
Grey thrush
Territorial
Snapping of beak of yellow robin
Territorial
Magpie
Territorial
Grey thrush
Crimson rosella
Kookaburra
Whipbird (terminal note)
Whipbird (full call)
Silver-eye
Grey thrush (shrike note)
Crimson rosella
Scissors grinding
Clonk clonk
Clickety clickety click
Yellow robin
Warrrh (a specific lyrebird note)
Snapping of beak of yellow robin
Clonk clonk
Scissors grinding
Yellow robin
Scrub wrens
Scissors grinding
Clonk clonk
Clickety clickety click
Grey thrush
Grey butcherbird
Scissors grinding
Yellow robin
Whipbird

TOTAL: 56 items in 180 seconds

This list of imitated calls may appear impressive, but this bird, at that particular time, did not imitate the pied currawong, gang-gang cockatoo, grey fantail, olive whistler, fantail cuckoo, brown-headed and crescent honeyeaters or the eastern shrike-tit, all of which are commonly heard in the mimicry of Sherbrooke's lyrebirds.

*An immature male lyrebird (Red/Blue) aged ten months, showing a bright rufous throat coloration.*

*A young male lyrebird, in his fourth year, showing only slight rufous throat coloration.*

*A mature male lyrebird growing a new tail following the loss of his sixteen tail feathers.*

*This bird had acquired eleven filamentaries during his eighth year, but still retained one plain feather. This feather, next to the left lyrate, is clearly barred and strongly pigmented similarly to the lyrate feather.*

*This picture shows the common fault bars in feathers, attributed to dietary deficiencies formed during the coldest part of the night.*

*Two immature male lyrebirds displaying on the same mound, providing mutual stimulation.*

*Spotty, on a moss-covered log, pauses in his song before running to a mound.*

# 9 | Lyrebirds in captivity

The idea of keeping lyrebirds in zoos might well have originated with Dr Ludwig Becker (1857), who took a six-week-old chick (and its mother, for a specimen), in 1854. He managed to keep the chick alive for eight days and remarked that lyrebirds could readily be tamed and that they would do well in zoos.

Apparently it was fashionable in those early days for bird fanciers to keep (or to attempt to keep) a few lyrebirds in captivity. Dr George Bennett, a Sydney doctor whose hobby was ornithology, wrote a book, *The Wanderings of a Naturalist in Australia* (1860), which attracted great interest. In a letter dated 27 March 1867 to Dr J.E. Grey, then President of the London Zoological Society, Dr Bennett referred to a Mr J.S. Palmer who, over a period of years, had endeavoured to rear lyrebirds of all ages, which he had obtained from the Illawarra District, south of Sydney. Unfortunately, all but one of Palmer's birds had died after a short period. The sole survivor, a young male, had been in custody for some two years and, even at the beginning of his third year, was already a competent songster. He is said to have imitated such birds as the wonga pigeon, black-backed magpie and other birds with which he shared the aviary. Dr Bennett had tried unsuccessfully to buy the bird for shipment to London.

An interesting story about the keeping of lyrebirds in captivity comes from J.C. Mahan (1905) who lived in the Woods Point district in Gippsland. Having found a nest, it was Mahan's practice to wait until the chick was about thirty days old before snaring the female in the nest, at night-time, and then transporting both chick and female to his large aviary. From this point on in his narrative, Mahan is less frank; he does not disclose details of the food he provided for the birds; but he left it to the experienced mother to feed the chick, instead of repeating Becker's mistake of attempting to rear the chick unaided. However, he was not eminently successful, because he sacrificed no fewer than thirteen chicks before he claimed to have been able to bring 'this tedious and difficult task to a successful conclusion'. LeSouef offered him £40 for his collection; but, later, he said, he was on the point of accepting an offer to supply three pairs of domesticated lyrebirds to a London zoo, at £115 a pair, when his entire stock was 'maliciously poisoned', whereupon he abandoned the project.

There are many others who have kept lyrebirds, but mention must be made of Mr and Mrs Jack Coyle of Springwood, in the Blue Mountains district of New South Wales. They succeeded in rearing two birds, taken from their nests as fledglings, which they named 'Joe' and 'Zoe' and which later achieved much publicity through being photographed in the company of many glamorous admirers.

In an account published in the *Argus* (28 June 1936), A.H. Chisholm reported that Zoe built her first nest at the age of four years, but did not lay an egg; in the following year she built again and laid her first egg, which she destroyed twelve days later. Information is lacking on her nesting activities over the next few years; but, in 1936, Zoe's ninth year, she and Joe combined to produce their first chick, which appears to be the first lyrebird to have been hatched in captivity. This chick, named 'Pat', thrived in the care of Mr and Mrs Coyle, but, through a series of unfortunate accidents, all three birds died, though not before Joe and

Zoe had completed a partnership of some fifteen years and demonstrated that lyrebirds could breed in captivity.

# Lyrebirds in zoos

It seems curious that, with its excellent facilities, Taronga Park Zoo in Sydney has not developed an attractive lyrebird exhibit. Several years ago, in response to my inquiries, the late Sir Edward Hallstrom informed me that 'many years ago, we had a female lyrebird which was with us for twenty years. She never built a nest or laid an egg. At present, we have two males, not yet fully mature'. Recent advice from the Zoo indicates that there has been no significant development in regard to lyrebirds since then.

The lyrebird appears never to have been exhibited at the Royal Melbourne Zoo. This may be because it was felt that Sherbrooke Forest, only 40 km distant, could provide all that was required in the way of a 'lyrebird spectacular' without the considerable expense of constructing and maintaining a lyrebird exhibit. The Healesville Sanctuary (formerly the Sir Colin McKenzie Sanctuary) is now affiliated with the RMZ, so it seems unlikely that, in the foreseeable future, there will be such an exhibit at the Zoo.

It would appear that the first zoo to capitalize on the attractions of the lyrebird was the London Zoo. A report by the Superintendent at the time, A.D. Bartlett (1867), stated that on 9 April 1867 a fine example of a superb lyrebird had been received at the Zoological Society's Gardens. It was a young bird of unknown sex; but, as it had been reared from the nest, it was used to people. The report does not say who the bird's foster parents in Australia were, how it was cared for when young or how it fared during the long sea voyage, but presumably there were few complications. In London, it was fed on a mixture of finely chopped meat, bruised hemp seed, earthworms, mealworms, ants' eggs and grasshoppers, together with a little canary and millet seed.

The young bird quickly adapted to its new environment and caused much interest by its speed of movement and by its ability to hop and spring 'incredible distances'. The bird was tame to the point of being friendly and would use its claws in a quietly persuasive manner to try to force open the hand of its keeper to obtain the food which the bird thought was there. (I have had similar experiences in the forest with young lyrebirds which have developed a liking for cheese.) Bartlett remarked on the great strength of the bird, which could lift a 7 lb (about 3 kg) clod of earth, and on its loud and powerful voice.

In May 1868 it was reported that a male lyrebird of the same species had arrived at the zoo on 21 April, 'making a pair of this extraordinary bird now living in the Society's Gardens'.

It is not known whether the Society received any further lyrebirds from Australia; but, in the *Argus* for 26 August 1936, there appeared a report by A.H. Chisholm to the effect that a lyrebird had died in a London zoo, in 1903, and was never replaced. Inquiries revealed that unfortunately many of the Society's records were lost during World War II, so the history of this bird and the others cannot be traced. From 1867 to 1903 is a very long time for a lyrebird to live, but it is obvious that the birds were well cared for and protected from predators, so that the possibility that a lyrebird could live for thirty-six to forty years should not be discounted.

Chisholm (1960) refers to the keeping of a lyrebird in a Paris zoo, in the 1880s, for a period of five years, but no other information appears to be available.

In 1949 two lyrebirds were introduced into the Adelaide Zoo, having been caught in the Healesville (Victoria) district by David Fleay, the well-known naturalist, who was then the Director of the Sir Colin McKenzie Sanctuary. At the time, it was thought that one bird was about eight months old and the other a year older, and that they comprised a pair. As events turned out, they were both females. When I first saw them in Adelaide in 1954, they both appeared to be healthy, despite the

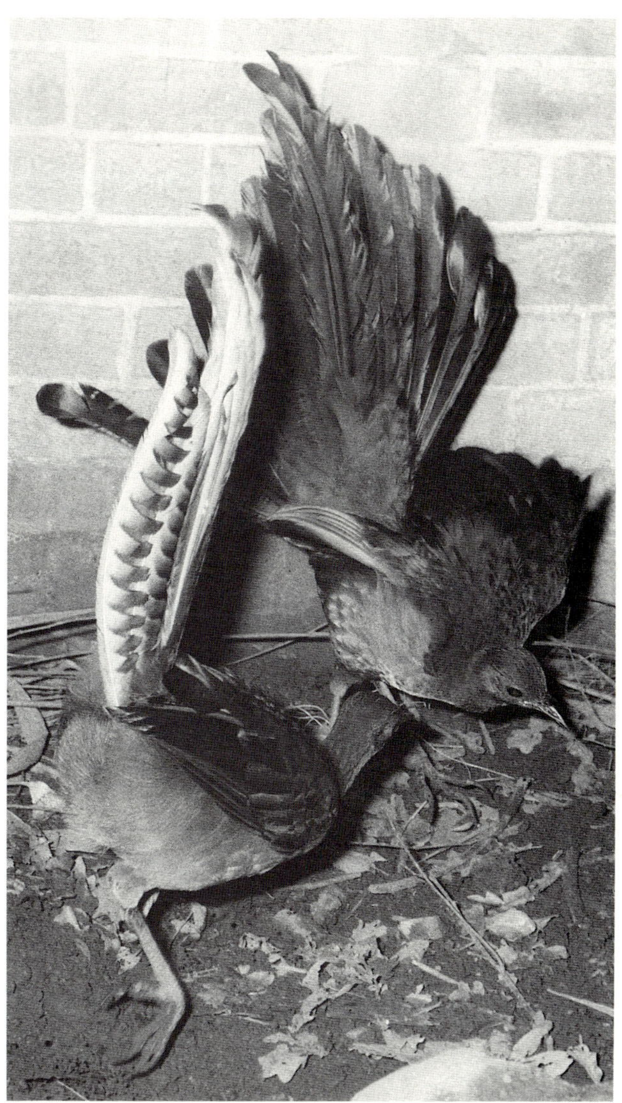

*Two female lyrebirds in the Adelaide Zoo, in a mutual display.*

small size of their enclosure. In due course, both birds built nests and laid eggs, but these were of course infertile. That two female lyrebirds should have built their nests a few feet apart suggests that, under adverse conditions, the drive to build a nest may supersede the bird's natural tendency to exercise its territorial rights and preference for seclusion when building its nest. In some cases, the eggs were taken by rats and, later, one of the birds took to eating its own egg. When I saw them in August 1967 they were both active and I was able to photograph them engaged in a mutual display. Unfortunately, both birds died about a month later and have not been replaced.

While the Healesville Sanctuary is not a zoo in the accepted sense of the word, it is nevertheless a place where animals and birds are kept within enclosures. The 'Sanctuary' grew out of the researches of the late Sir Colin McKenzie on certain Australian animals, including the koala and the platypus, and was established in 1929, when Sir Colin moved to Canberra to establish and direct the Australian Institute of Anatomy. When I first visited the Sanctuary in March 1933 it was indeed a place of modest pretensions and there were no lyrebirds present. It was not until January 1939, when a male lyrebird is said to have sought refuge from the bushfires which were then raging in the district, that the Sanctuary acquired its first lyrebird.

In the years that followed, the Sanctuary has provided a home for a number of females or 'plaintails', but was less successful with male lyrebirds, apparently because of the lack of a suitable enclosure.

In 1963 the Royal Automobile Club of Victoria launched an appeal among its members and raised the sum of $A31,600, to which the Victorian government added $A13,000, to enable a large aviary, said to be the largest in the world at the time, to be constructed. The aviary measures 65 m long by 33 m wide, the sides being 15 m high. It was designed to permit as much as possible of the natural vegetation to be enclosed and the environment was developed within to create the effect of a 'little Sherbrooke'. The aviary was officially opened on 15 April 1965 by the state Governor, Sir Rohan Delacombe.

The first occupants were a male and female; the latter, in 1965 and 1966, built a nest and laid an egg which was infertile. Since that time, several chicks have been hatched and have thrived for several years. However, there have been at least two casualties due to the aggressive behaviour of a brush turkey which was housed in an enclosure adjoining that to which the young lyrebirds had been transferred from the main enclosure.

In the RACV enclosure, visitors are able to walk through a miniature forest and observe the lyrebirds engaged in their various activities, including the display of the male. At long last, also, the opportunity exists for research on the lyrebird, though not at the Royal Melbourne Zoo, as I had once hoped it might. Such work should supplement the excellent fieldwork by the Sherbrooke Survey Group and others.

# Our tame lyrebird

It was never my ambition to conduct a private zoo, but the circumstances which so unexpectedly made us the guardians of a mature male lyrebird, and the subsequent events, now have an important place in this book.

On Friday, 15 August 1969 I addressed the members of the North-eastern District Fly-fishermen's Association, at Wangaratta, in north-eastern Victoria. Norman Gibb, the then President of the Association, acted as host to my wife and me during our visit and, on the following day, took us to the environs of Mount Cobbler, which some people saw as a potential national park. On Sunday, we were taken to Black Mountain, where we were introduced to the local lyrebirds. These birds have a distinctive territorial call, and their song differs in many respects from that of the lyrebirds of Sherbrooke.

After an eventful day in the field, we were enjoying dinner with our hosts, when the telephone rang. The caller was Pat Sherrett, Fisheries and Wildlife Officer for the district.

*Our 'tame' lyrebird, taking a drink of water from a glass held by my wife.*

My surprise can hardly be imagined when he asked me if I would be prepared to look after a lyrebird which had been delivered into his care earlier in the day. The bird had been struck by a car on the Mount Buffalo Road several hours previously. A driver who had witnessed the accident rescued the lyrebird from the gutter at the side of the road and, although the bird appeared to have had a lucky escape, decided to take it to the FWO so that it could receive any necessary attention. Pat Sherrett, full of compassion for the lyrebird, thus found himself in a dilemma, because he had had no experience with lyrebirds. However, he had met me on the Friday evening and knew of my interest in the subject; hence the call.

Thus it came about that, on Monday, 18 August 1969, I found myself driving back to Melbourne in the company of a live mature male lyrebird in a large cage which occupied the whole of the back seat of the car. The bird seemed very tractable and in good health; there was no sign of blood on his beak, but he had been without food and water for twenty-four hours. I was anxious to make good these deficiencies; so, as soon as we reached our home in Kew, we set about persuading him to eat and drink. Although he stood quietly, he showed no enthusiasm for our offerings. I poured water into my cupped hand and gently pressed his beak into it, without success. I ran my wet fingers along his beak, hoping to squeeze a little water into it, but all to no avail. Then I opened his beak and allowed a few drops of water to run off my middle finger into his mouth ... it all seemed so hopeless. But then, suddenly, he began to move his throat muscles as if tasting the water. I offered him my cupped hand again and was overjoyed when he voluntarily moved his head towards my hand and took a sip, then held his head up to allow the water to run down his throat. Soon, he was back for more and shortly we had him drinking out of a glass, as if that had always been his preferred method. Round 1 to patience and perseverance.

Next we had to induce the bird to eat. First we tried him with worms freshly dug from the garden, but he seemed to have forgotten how good worms can taste (to lyrebirds). I opened his beak and placed a worm in it; he just stood there with the worm dangling from his beak, but made no effort to swallow. Eventually, the worm wriggled free; surely this must have been the only worm ever to have escaped from a lyrebird. We tried to interest him in bread crumbs and tiny bits of cheese. These are like manna from Heaven to the lyrebirds of Sherbrooke, but this bird showed no interest. In desperation, and to give myself a little respite, I stood the lyrebird on a small log and tossed a worm on to it. The bird looked at it cautiously, as if it had stirred some memory of the distant past, but then the worm began to wriggle in an effort to escape.

That proved to be his undoing; in a flash, the lyrebird picked up the worm and swallowed it ... and likewise with several others.

When we had given him a good meal of worms and more water, there were other things to think about. Where were we going to keep him? I wanted him to be under close observation, because I needed to be sure that he had recovered from his encounter with the car. I had been informed that he had struck it a glancing blow, which was much better than a direct hit. So we spread newspapers all over the floor of the sun-room at the northern end of the house and likewise protected the couch on which my wife sometimes basks in the midday sun. We had been successful a few years previously in house-training a baby wombat, but felt less optimistic about our chances with a lyrebird.

As I had no information on the weight of a mature lyrebird, I saw this as a great opportunity; so we weighed him on the kitchen scales. At 852 g he stood on the scale pan while his photograph was taken; there was no fuss, even when we approached him, nor did any sound escape from this Prince of Melody.

The bird stood on the couch, gazing out through the large window into the garden below, and I wondered if he was remembering the vast expanse of his forest home, now so far away. We decided to take him out into the garden; he followed us around quietly as we dug for worms, accepting our offerings but making no attempt to scratch.

At the end of that first day, we felt that we were making progress ... the bird was still alive, and was eating and drinking, showing no sign of distress as he walked about. When night came, we had to protect him from chance molestation from marauding cats which sometimes prowl about our garden. We don't have a dog or cat. We decided that he had better sleep in the bathroom; so, having protected the floor with newspapers, we presented him with a small log beneath the wash basin, and there he perched. Soon he drew one leg up into his feathers and tucked his head under one wing. These seemingly quaint

habits are designed to reduce heat losses from lightly or unfeathered areas (Welty, 1982). In the morning, while I was shaving, the lyrebird remained on his perch, between my feet, before being taken outside to be fed and watered.

This was the daily routine for the remainder of the week. Of course, I had to be away from home at my office during the day, but my wife dug for worms for the lyrebird and I did likewise after arriving home in the evening. The bird seemed quite happy as he moved about the garden, which was rapidly being dug 'up and over'. At the end of the week, the lyrebird's weight was 907 g and the prospects for a successful outcome seemed hopeful. Hopeful for what? I was beginning to ask myself; we could not offer him a permanent home.

Away back in 1956, at the request of the then Director of the Royal Melbourne Zoo, I had prepared a submission for consideration

*Our lyrebird being weighed on the kitchen scales. This picture was taken several days after the first weighing and shows that the bird had gained 55 g in weight.*

by the Scientific Committee, with the object of establishing a large aviary in which a few lyrebirds could be exhibited to the public, in a miniature Sherbrooke Forest setting in which other compatible species of Australian birds could also be shown. My object was not merely to exhibit the lyrebird, but principally to provide an opportunity for scientific studies on Australia's most famous bird. To date, there had been no really scientific study of the species; I was virtually at the beginning of my work on the development of the tail feathers and the sequence of moults and, at that time, of course, could not foresee the outcome. My plans were considered attractive, but the proposal lapsed for lack of finance.

But now I had a live lyrebird to offer the Zoo Board. I decided to discuss the matter with Mr A. Dunbavin Butcher, who was then the Director of Fisheries and Wildlife and Chairman of the Royal Melbourne Zoo Board. He was also one of my colleagues on the National Parks Authority. Mr Butcher was enthusiastic and encouraging, although we both recognized that we had many hurdles yet to jump. But at last I thought I saw a glimmer of light at the end of my long tunnel.

On Saturday afternoon, 24 August, our daughter Helen paid us a visit to meet the new member of the family. The lyrebird perched obligingly on the back of her chair while I took their photographs. Helen was delighted, but somewhat startled when the lyrebird made a lunge at her tongue as she laughed, obviously mistaking it for a large worm. And, on the following day, the lyrebird behaved graciously for my neighbour whose two young sons played with the bird on the lawn while their father made a movie film of the group.

On Monday morning, just a week after I first met him, the lyrebird was given his breakfast and water, and seemed to be in good health, but, when I arrived home from work, I found him dead; no longer was he a lyrebird in captivity. Words cannot describe my feelings of sadness and frustration at the loss of what had become a valued friend. I wondered whether, if I had taken the bird to a veterinary surgeon, he might have given it a shot of antibiotics to neutralize any infection resulting from the accident.

With my dreams and hopes dissipated, I handed my dead lyrebird over to the Fisheries and Wildlife laboratory in the hope of learning the cause of death; the verdict was 'delayed shock'. I kept the tail feathers of the dead bird, and these proved invaluable to me in my studies on the lyrebird's tail feathers, as described in chapter 7. Thus ended one of the most interesting incidents in my life with the lyrebird.

# 10 | Lyrebirds in Tasmania

Although the lyrebird is now present in considerable numbers in central and southern Tasmania, it is not an indigenous species. What appears to be a plausible explanation of the absence of lyrebirds from Tasmania and Wilsons Promontory has been given in chapter 2. Had the link between Wilsons Promontory and the mainland which eventually formed been a suitable habitat, there is little doubt that the lyrebird would have colonized the Promontory, because they were originally present in good numbers in the Fish Creek district, before land clearing and settlement dispersed them. Fish Creek is only about 40–50 km 'north' of the Promontory. Information supplied by Mrs Ellen Lyndon of Leongatha indicates that, even as recently as a decade ago, there was still a small colony of lyrebirds in the Walkerville district, which lies to the south of Fish Creek, but whether they still persist there is not known. Between 1910 and 1912 two male and five female lyrebirds were transferred from the Fish Creek area to Wilsons Promontory, but it appears that the species is now extinct there, although the mountain gullies would appear to be ideal habitat.

Paradoxically, perhaps, it was the fact that Tasmania had become an island, isolated from the impact of some of the destructive changes wrought by man on the mainland, which engendered the idea of introducing the lyrebird to the island state. The wholesale slaughter of the lyrebird which had characterized the first hundred years and more of white settlement and the loss of habitat as land-clearing for farming and other human purposes proceeded, and the increasing menace of the introduced fox, gave rise to a growing concern for the future of the lyrebird. 'The days of the lyrebird are numbered', wrote A.E. Kitson in 1905, 'unless . . .'.

In 1908 this note appeared in the *Emu*:

> The following letter to the Hon. the Chief Secretary, Hobart, has been forwarded to Mr A.E. Elliott, Hon. Sec. of the Tasmanian Field Naturalists Club: "I have the honour to inform you that I have been delegated by the Australian Ornithologists Union to bring to your notice the desirability of introducing lyrebirds (*Menura* sp.) into Tasmania from Victoria. Owing to the destruction of these birds by foxes in the mainland, they are threatened with extermination and, in order to prevent this, it is proposed to import some into Tasmania where the deep mountain gullies and fern glens make ideal resorts for these unique birds."

In the following years, the proposal was kept alive by exchanges of correspondence between the relevant Victorian and Tasmanian departments and by the enthusiasm of a small band of conservationists. There was, of course, some opposition; it was argued that such introductions would disturb the balance of Nature (as if Nature was not already reeling, utterly unbalanced, through the impact of white settlement with its machinery of destruction, the fire-stick and introduced pests). But the plan began to take shape; sponsors were found who were willing to contribute money to hire those who would need to be engaged to catch the lyrebirds.

One obstacle appears to have been the fear that the lyrebirds might not survive the sea-crossing from Melbourne to Tasmania. Apparently it was overlooked that, as far back as the 1860s, lyrebirds had survived the long voyage from Australia to Europe, where they took up residence in zoos in London and Paris.

The plan received a great impetus in 1934, when Holyman's Airways, the pioneer air service across Bass Strait and the forerunner of Australian National Airways, offered free transport for the lyrebirds across the Strait. Following this, there was a road journey of about 230 km, the last 6.5 km by packhorse. Mount Field National Park was selected as having ideal lyrebird habitat; the authorities were aware of the difficulties and were taking all precautions.

The first two lyrebirds, a male and a female, were caught by Harry Howe, of Canterbury, Victoria, using special spring snares, in the forest near Selby, not far from Sherbrooke Forest, 'without the loss of a single feather'. Photographs of the birds, along with their captor, which appeared in the daily papers of 27 August 1934, suggest that the birds were in good fettle. The male bird actually gave a fine concert in his cage, while waiting for the train to Melbourne, before the air trip.

The two birds arrived in Tasmania on 29 August and were immediately taken to the Park, where they were released. Next morning, the male bird was found dead, but the female was reported well. The death of the male bird brought strong protests from lyrebird lovers in the Dandenongs, but the Royal Australian Ornithologists Union expressed its support, while ministers of the Crown and government departments rallied to the cause.

However, there was clearly no intention of flooding Tasmania with avian immigrants, and a year was to pass before three additional males and two females were released in the national park in August–September 1935. Thus, by the end of 1935, three female and four male lyrebirds (of which two are known to have died) had been released in the Park.

The following table gives the full list of lyrebirds transferred to Tasmania over a period of fifteen years. The first eleven birds were caught by Harry Howe in the Selby area; the remaining eleven birds were caught by David Fleay (1952) in the Toolangi–Warburton district, some 60 to 80 km east of Melbourne.

Considering that a successful mating can produce only one chick from each pair of lyrebirds each year, which, on mainland statistics, has much less than 100 per cent chance of survival, the scale of operations can hardly be said to have been lavish. It will be interesting, I hope, to see how the experiment turned out.

My first experience with the lyrebird in Tasmania occurred while I was attending a conference there in July 1941. Our hosts, Australian Newsprint Mills, gave the visitors the pleasure of inspecting logging operations in the Florentine Valley and of visiting the beautiful Russell Falls, near the area where the first release of the lyrebird had been made about seven years earlier. I was excited at finding the unmistakable scratchings and droppings of the lyrebird near the Falls, giving proof that the lyrebirds were not far away. However, we did not make the acquaintance of the lyrebird in Tasmania on that day and I had to wait a further thirty-nine years before that wish was gratified.

Reports on the small lyrebird colony were, of course, awaited with great interest; but in those early days information was sparse. Michael Sharland, who had a long-standing association with lyrebirds (having visited Tom Tregellas in his log home in 1929–31), was a pioneer in this field. Sharland, a noted ornithologist and photographer, and a superintendent of the Scenery Preservation Board in Tasmania for many years, found the energy and enthusiasm to pursue the matter and provide vital information.

Sharland reported in the *Emu* (1944) that several records were available to show that a lyrebird had been seen near the area where the birds had first been released in the national park. Such sightings referred mostly to a male; there were few sightings of females on record, and none regarding the discovery of a nest. The birds had not all remained near the area of release; Gordon Rowe stated that on 7 February 1944 he had seen a mature male near Wherrett's Look-out, south-west of Mount Field, some 32 km from where the birds had

## Lyrebirds transferred from Victoria to Tasmania

| | | |
|---|---|---|
| 28 August 1934 | 1 male<br>1 female | Released in National Park 29 August; male found dead next day, female well |
| 14 August 1935 | 1 male | Found dead on 22 August |
| 23 August 1935 | 1 male | Released in National Park |
| 3 September 1935 | 1 male<br>2 females | "   "   "   "<br>"   "   "   " |
| 26 August 1938 | 2 males<br>2 females | "   "   "   "<br>"   "   "   " |
| 5 November 1941 | 1 male<br>2 females | "   "   "   "<br>"   "   "   " |
| 26 January 1945 | 1 male<br>1 female | Released in Hastings Caves Scenic Reserve<br>"   "   "   "   "   " |
| 28 May 1945 | 2 males<br>1 female | "   "   "   "   "   "<br>"   "   "   "   "   " |
| 30 May 1945 | 1 female | "   "   "   "   "   " |
| 30 November 1949 | 1 female | Released in National Park |
| 7 December 1949 | 1 male | Released in National Park |
| TOTAL | 11 males + 11 females | |

been liberated. Early in the day, Rowe had flushed a male lyrebird and, a few hours later, he had heard it singing strongly. He was able to recognize a number of imitatory calls, including the mainland kookaburra. Scratchings indicated that the bird was active in that area, roughly 37 km from the release area.

Another report came from the park ranger that, in June 1943, a bird had been seen near the road which leads from the park entrance to the highlands, a considerable distance from the point of release. An unconfirmed report stated that some osmiridium miners had seen a bird on the track to Adamsfield, in June 1943.

In June 1941, a party consisting of Sharland, J.A. Tubb and H. Belcher spent a day combing the forest where the birds were first released; but they neither heard nor saw lyrebirds. Undaunted, Sharland, in June 1943, camped on the fringe of the forest, not far from the tongues of snow which stretched downwards from the highlands. Yet, although he heard several mature lyrebirds calling, and saw heavy scratchings, he failed again to see a lyrebird (Sharland, 1944). But his patience was rewarded in June 1945, when he found three old nests in a fern gully about 1.5 km from the point of release (Sharland, 1952). One nest contained an egg which, however, was empty and shattered when touched. A year later, Sharland found the foundations of a nest which, unfortunately, was not completed. He wrote that 'curiously, none of the birds released in the Hastings Caves area has since been seen'.

'All that can be said at the moment [wrote Sharland] is that, of the nineteen birds introduced, three pairs have become established in and adjacent to Mount Field National Park … One of these has its territory within the park, in beech forest, between the five- and six-mile posts along the road leading from the park entrance to the highlands; another pair has selected territory near Crisp's Hut on the Adamsfield track along the southern boundary of the park, and the third pair is at Risby's Basin, about five miles south of the park.'

Another pioneer in the Tasmanian lyrebird story is L.E. Wall, who in May 1950 heard a bird calling in the vicinity of the five-mile post along the road from the park entrance to Lake Dobson, obviously close to the area referred to by Sharland in 1952.

Wall has since paid numerous visits to this area and found that there was a greater concentration of lyrebirds there than in any other area he had explored. He was usually able to hear several birds calling at once and located some fourteen display mounds within a radius of about 180 m, many of the mounds having been in use for several years.

Wall had his first sighting of a lyrebird in July 1956; since then he has seen the male bird on several occasions, but did not see a female or an immature bird until 20 September 1956, when the ranger escorted him to a nest containing a healthy chick. This was the first such nest to be found since the lyrebird was introduced some twenty years earlier, proving beyond all doubt that the lyrebird was successfully breeding in Tasmania. It is presumed that this was the nest found in Marriott's Falls Scenic Reserve, about three kilometres west of the park boundary.

Wall made a number of recordings of the lyrebird's song, which enabled him and W. Roy Wheeler (1966) to compare the song of the Tasmanian lyrebird with that of Sherbrooke Forest. Because there was no way of knowing whether the birds recorded were related to those which originally came from the Selby district (near Sherbrooke) or whether they had their origins in the Toolangi–Warburton district, a direct comparison is not feasible, but this was the first attempt to analyse the song of the Tasmanian lyrebird. The two songs, not unexpectedly, had little in common; but, surprisingly, the call of the whipbird, which does not occur in Tasmania, was clearly heard in the song of the Mount Field lyrebirds.

Wall's recording apparently did not contain the call of the kookaburra, which had been heard by Gordon Rowe in 1944, in an area some 32 km from where Wall made his recording many years later.

It can be seen that, by the middle of 1966, that is, over thirty years after the first introductions were made, the lyrebird was well established in Tasmania.

# The Tasmanian experience

I was not able to revisit Tasmania until after I retired in 1975; but, towards the end of April 1980, I was able to resume my interest in the lyrebird in Tasmania. In preparation for the planned visit, I was fortunate in securing the wholehearted support of my old friend Peter Murrell, Director of the Tasmanian National Parks and Wildlife Service, and his staff, and of Michael Sharland and Leonard Wall. Mr Sharland provided me with a map, indicating where he had his successes, while Mr Wall gave useful information, tips on recording and promised to act as my guide. It seemed that all I had to do was to get myself and my recording equipment to the scene of action; but I allowed a period of five weeks, as a form of insurance against bad weather and other frustrations.

Early on the morning of the day after our arrival in April 1980 at the well-appointed camping ground of Mount Field National Park, my wife and I heard the familiar 'clonk clonk' call of the lyrebird from the forest, but we were isolated from the songster by the swift-flowing Tyenna River. On the advice of the Ranger-in-Charge, Trevor Westron, we were soon on our way to the Nature Trail, about eight kilometres from the camp ground, a little beyond the five-mile post which was now firmly fixed in our minds. On the trail, we renewed our association with the Antarctic beech and the magnificent mountain ash, as we walked wide-eyed through this cool, silent, beautiful forest. The serene quiet of the forest, basking in dappled sunlight, was soon broken by the staccato calls of a lyrebird … and then another. We felt that we were indeed in the Promised Land; but, after struggling through the dense forest, with huge fallen trees blocking our progress at every turn, carrying a 60-cm parabolic reflector on its heavy tripod, along with an Uher tape-recorder and microphone, we began to wonder whether our objective was to be easily attained, after all.

So we abandoned the chase and returned to the trail, only to be drawn away from it again across the road into the beech forest. Here we caught up with our quarry, a magnificent male lyrebird which, as far as we could see from a distance of 10–12 m, resembled the lyrebird of Sherbrooke. He was scratching quietly among the beech litter; but, from time to time, he disappeared behind a log, affording us an opportunity of creeping a little closer. If he were aware of my presence, he showed no concern and, when he sang for a short period, I made a brief recording; but soon he wandered off into the virtual jungle, leaving us alone.

The remainder of the day was spent coming to terms with the forest, making short excursions into it, learning to recognize useful landmarks to facilitate our return to the car and occasionally pitting myself against the horizontal scrub. An essential preliminary step in our task was to learn to identify the various calls of the native birds. This took time and would not have been possible without the invaluable help of the park staff and Len Wall. We were greatly intrigued by a kind of duet between an olive whistler and another bird which we were unable to identify. Had this bird been in Sherbrooke Forest, it would without hesitation have been recognized as a pilot bird, but the records show that this bird is not found in Tasmania. During our visit, these birds were frequently heard calling to one another over an area about 200 m in radius, in the vicinity of the beech forest near the Nature Trail. The piping note of the unknown bird was immediately answered by the familiar and unmistakable note of the olive whistler, and this continued for half an hour at a session, as the birds followed one another through the forest.

Our task was made much easier when, a few days after our arrival, Mr Wall came up from Hobart and introduced us to an area where he had had notable success in recording the lyrebird. In effect, the area where we were to work during the next few weeks was a large spaghnum moss bed (which becomes frozen in winter) where large pandanni plants mingle with silver banksia, eucalypts (notably *E. johnstoni*), celery-top pine, cutting grass (*Ghania* sp.) and other species. This area was separated from the beech forest by a dense belt of horizontal scrub. In the course of our visit, we found in excess of a dozen active mounds, mostly in the seclusion of the scrub, but some mounds had been formed in the moss bed itself, being surrounded by celery-top pine and cutting grass. The beauty of the setting was enhanced by coral lichen, a plant which fascinated us with its delicate structure. There were two mounds at the edge of the forest, where we parked the car, close to the entrance to the Nature Trail. We soon learned to identify some six singing males around the periphery of the Pandanni Flat; frequently there were several singing at the same time. Interspersed within the moss bed, which was decidedly springy underfoot, were several crystal-clear pools in which the surrounding forest was reflected to perfection.

Cautiously, we familiarized ourselves with the area and memorized the landmarks which we hoped would lead us swiftly to the proximity of the several mounds we had selected as likely to yield recordings. Our plan was to sit and wait quietly for a lyrebird to call, then to proceed as quickly and as silently as possible to a spot deemed suitable for recording, set up the tripod and parabolic reflector, and record. The use of the parabolic reflector, of course, rendered it unnecessary to be closer than 20 m; but, as we found to our cost, the birds have very acute hearing and sometimes left the mound just as we were about to commence recording. Nothing comes easily in Lyrebird Land.

While we waited for the lyrebird to call us to arms, we were regularly entertained by the delightful 'rolling' notes and the staccato 'chop chop' call of the yellow-throated honeyeater and the resounding 'Egypt' call of the crescent honeyeater, which fed together in the trees above and around us. Occasionally, the excited chatter of the black-headed honeyeater reminded us of the brown-headed honeyeater of the mainland, and the small fry such as the scrub wrens and scrub tits seemed always to be talking to one another. From the forest-at-large came the loud screech of the yellow-tailed black cockatoo, the rather quaint notes of the green rosella, the plaintive notes of the black currawong and the song of the grey shrike thrush, though this, we thought, lacked the quality and volume of the mainland grey thrush. We never felt alone as we waited. Our greatest concern was the wind, which played boisterously and almost incessantly among the tops of the trees which covered the slopes around us, providing an unwelcome background noise in our recordings.

One afternoon, I followed a lyrebird from one mound to another, in a desperate effort to obtain a recording of a fine songster, but permitted my enthusiasm to carry me too far into unfamiliar territory and, when I 'turned for home', the sun had gone behind dense cloud and I was unable to locate the faint trail which would have led me back to the road. For several hours I sought to penetrate the horizontal scrub while the light paled, and eventually I realized that I was in a serious predicament. My wife, who had been sitting in the car awaiting my return, had become anxious and sounded the horn but failed to hear my reply. She walked down the road to the place where she had seen me enter the forest and called to me, but again failed to hear my reply. I endeavoured to push through the scrub to the point from which she had called, but to no avail.

Time and again, I found myself forced to turn, first to the left, then to the right, then to the left, in order to get around some obstacle and escape from the horizontal scrub and dense cutting grass. Then I slipped into a depression from which I could not extricate myself. I realized that I might have to spend the night there. When darkness descended like a pall on the forest and the stars began to shine brightly (the clouds having for a while dispersed), and the owls began to hoot, and the sweat in my clothes became cold and clammy in the freezing air, I wondered whether I could survive the rigours of the night.

But, fortunately, the park ranger had noticed that we had not returned to the campground and set out to find me. Just when I had abandoned all hope of being rescued that night, I heard a vehicle coming up the road and saw the headlights. Soon we were exchanging greetings and I could hear the ranger and his assistants as they strove to reach me. They had great difficulty in negotiating the numerous obstacles which Nature had placed between us; but, after what seemed an eternity, I felt the warm hand of the ranger pulling me out of my trap and shortly I was on my way to safety.

Although the kindly ranger said not a word of reproach, I felt chastened and resolved not to repeat my mistake. But, even in my hour of remorse, I felt a tiny glow of satisfaction, because, on that day, I had made my first good recording before losing my way. After that, we marked the trails with red tape, so as to facilitate our return to safety.

We spent almost two weeks at Pandanni Flat; but, of course, the birds did not sing every day and the weather was not always favourable. However, we gained not only several good recordings, but also a familiarity with the calls of the other birds of the area, which enabled us to interpret the lyrebird's mimicry.

Although the lyrebirds could be heard singing on most days, at almost any time after we arrived at Pandanni Flat (about 8.30–9 a.m.) until we departed (4–4.30 p.m.), they were usually at their best (obviously on their mounds) at about 11 a.m. to 12.30 p.m. and again between 2 p.m. and 4 p.m. The duration of the song varied from 10–15 minutes to 30 minutes and longer. One of the best performances was that given by a bird which was already singing on a mound when we arrived at nine one morning and continued to sing for some 40 minutes. Having located the mound

some days previously and selected a well-screened spot from which to record, about 15 m away from the mound, I was able to obtain a good recording, marred only by the strong wind which was raging in the higher parts of the forest.

While we were at Mount Field National Park, we visited several other areas where lyrebirds were reported to be present in good numbers. At Ellendale, Keith Clark took us to a lightly timbered ridge above his property, where numerous mounds attested to the presence of lyrebirds, although we did not hear any singing at that time. Mr Clark informed us that, in the springtime, the lyrebirds called incessantly throughout the day. On a second visit to this district, we went in the opposite direction to Clem Dillon's farm near the forest. Even before we were out of the car, we could hear the familiar 'clonk clonk' call from the forest, about half a mile away. We walked along a track through fern gullies flanked by eucalypts, olearias, pomaderris, musk, blanket-leaf, sassafras and other species and eventually came close to the male lyrebird which had greeted us on our arrival. He moved away, but a female ran across the track in front of us, uttering the familiar alarm call. We heard several other lyrebirds in this area, but the rugged nature of the terrain deterred us from pursuing them.

Through the kindness of Gerard Cross, superintendent of Australian Newsprint Mills at Maydena, the centre of the company's logging operations in the district, we were guided up the Florentine Valley to Misery Plateau, and soon heard lyrebirds calling from the dense forest. The marginal vegetation defied entry, but I was fortunate in finding an old snig track, a relic of the logging operations of earlier days, and cautiously walked along this until I was opposite a fine songster, where I made a good recording. This bird appeared to be on a mound, with another bird close by, which interjected from time to time, as the song continued for half an hour

During the period 7–13 May, we stayed at an excellent cabin provided by the Ida Bay Railway Company at Lune River, about 5.5 km from Hastings Caves. This beautiful scenic reserve, with its wealth of rain-forest vegetation, provides an excellent home for the lyrebirds. It seems hardly likely that any of the six lyrebirds introduced in 1949 remain (though this is not certain), but certainly their progeny give ample evidence of their enthusiasm for the area. On our first visit, late on the afternoon of 7 May, we had walked only a few paces along the track to the Caves when the loud ringing notes of a lyrebird welcomed us.

We spent several days in the vicinity of Hastings Caves and soon learned how to penetrate the scrub. While in the Caves area, we visited adjoining parts of the forest, three to four kilometres from the Caves, and heard lyrebirds calling in every case. However, it was not in the dense forest that we made our best recordings.

At midday, 8 May, I was waiting patiently near the track, gazing at a tall stump which had been preserved to demonstrate the early logging methods, the platform on which the men stood, and the large cross-cut saw still remaining in position, as if the men had just knocked off for lunch. As I waited, an immature male lyrebird slowly made his way down the fern gully, warbling the while, and settled down about 20 m from the recorder. Apparently he had located a good feeding patch, because he was there for some 20 minutes, singing intermittently, before crossing the track and losing himself in the fern gully. During this brief interlude I resorted to a trick I had often played on lyrebirds in Victoria. I began to whistle, just as a stockman whistles up his dog.

Two days later, I was in a different area, a short distance from the logging exhibit, when a lyrebird began to sing from a mound on the slope above. Although I was unable to approach the bird as closely as I would have liked, I nevertheless obtained a creditable recording, from a distance of about 70 m. After the bird ceased singing, I remained where I was, some distance from the track, within the forest. After a short while, I heard the lyrebird moving just below the ridge, towards the fern gully below. I remained motionless (and, I

hoped, out of sight), but prepared to record, in case the bird should burst into song. When the bird was about 20 m away from me, he suddenly became aware that he was not alone. Had I moved, he would have taken flight immediately; as it was, he was perplexed, but not alarmed. He began to chortle, producing a medley of sounds which did not seem to be related to either his mimicry or his specific lyrebird calls. This is typical distraction behaviour of lyrebirds which I have often heard on the mainland.

As I remained motionless, he continued slowly to walk through the scrub, until he was less than two metres away from me. Until then I had not seen him; he was a magnificent male. There was a large log between us, but I was clearly visible to him. I don't suppose he had ever before come face to face with a man clad in green clothes and partly obscured by a 60-cm Grampian parabolic reflector equipped with an Uher microphone connected to an Uher tape-recorder. He soon showed what he thought of this intrusion into his private domain; he hissed, squawked, chortled, walked back and forth, swishing his tail angrily, scratched in a desultory fashion (again, typical distraction behaviour), and then turned and walked slowly back up the hill, obviously unhappy, puzzled by my presence yet unable to comprehend this seemingly immovable object. He came back to the other side of the log and resumed his hissing, squawking, chortling, even warbling occasionally. This performance went on for half an hour; it was not easy to maintain my pose, especially when I was assailed by mosquitoes one of which found the focal point of the parabola.

I felt richly rewarded when I heard the lyrebird giving a perfect imitation of my own whistling. This indicates the ease with which lyrebirds learn to imitate a sound which appeals to them. I have often wondered whether this item has found a permanent place in his repertoire and whether other lyrebirds will pick it up from him. The idea is not utterly unrealistic, because one may still hear the lyrebirds in Tasmania, in the several districts

*A view of Pandanni Flat, named after the pandanni* (Richea scoparia) *which flourishes there.*

*A lyrebird mound on a bed of spaghnum moss, surrounded by cutting grass* (Gahnia grandis), *celery-top pine* (Phyllocladus asplaniifolius), *etc.*

*The lyrebird imitates many other birds, including the species pictured here: the kookaburra.*

*The crimson rosella.*

*The golden whistler (male).*

*The grey thrush.*

*The red wattle-bird.*

*The yellow robin.*

*The satin bower-bird at his bower.*

*The grey fantail.*

*A copy of Gould's original plate of the Albert lyrebird, courtesy of the State Library of Victoria.*

we visited, imitating the whipbird, although this bird is not found in Tasmania. The whipbird's notes have been passed from the original lyrebird stock to their descendants and, although there has been some loss in quality, they are still unmistakably recognizable.

My natural modesty prevents me from saying how highly gratified I would feel if a small part of me could be immortalized in the song of the lyrebirds of Tasmania! Apparently, whistling notes sound pleasing to the lyrebird's ear, because L.C. Cook described a lyrebird's imitation of a stockman whistling up his dog in the Poowing district in Gippsland, in 1916.

# Growth of the lyrebird population in Tasmania

The object of introducing lyrebirds into Tasmania was to ensure the future of the species which seemed threatened with extinction on the mainland, so it may be interesting to examine the prospects of success.

As mentioned in an earlier chapter, the usual breeding cycle runs something like this: courtship and mating take place in May–June; eggs are laid in June–July; the chicks hatch in August; the chick leaves the nest late September–early October.

As the first two lyrebirds did not arrive in Tasmania until August 1934 and as the male died shortly afterwards, the female was unable to breed in that season. Of the two males introduced in 1935, one died and it was probably too late for breeding in that year, even if the 1934 female had been able to find the 1935 male survivor, but they *might* have begun to breed in the winter of 1936. As a result of the introduction of one male and two females in September 1935, at the beginning of the 1936 breeding season, there would have been two males and three females. If we assume polygamous union(s) (see page 24),

there could have been three chicks produced in 1936.

Now, studies on mainland lyrebirds have shown that the male matures at the age of seven to nine years and the female may mature in her seventh to ninth year. I shall assume that birds of both sexes mature in their eighth years. If two of the chicks produced in 1936 were females, they could have reared two chicks in 1944. If, however, only one female chick had been produced in 1936, there would have been only one chick in 1944 resulting from the progeny of the first breeding females which were introduced in 1934 and 1935. The progeny of the two pairs introduced in 1938 would not have commenced breeding until 1947.

The maximum number of chicks which could be produced in any season is governed by the number of breeding females available at that time. Thus, at the onset of the 1944 breeding season, the lyrebird population in Mount Field National Park *could* have been as follows:

| | |
|---|---|
| Original 1936 stock (3F + 2M) | = 5 mature birds |
| Chicks from 1936 stock | = 24 immature birds |
| Original 1938 stock (2F + 2M) | = 4 mature birds |
| Chicks from 1938 stock | = 10 immature birds |
| Original 1941 stock (2F + 1M) | = 3 mature birds |
| Chicks from 1941 stock | = 4 immature birds |
| TOTAL = 50 (12 mature + 38 immature birds) | |

Of the twenty-four birds produced by the 1936 stock, three would mature in the winter of 1944, as would three more in each succeeding year until the entire twenty-four had matured; but, in the absence of information regarding the sex of such birds, we need not speculate further. Indeed, Michael Sharland's discovery of three nests in 1945, one of which contained an addled egg, shows that our assumptions are too optimistic.

The two birds introduced into Mount

Field National Park in 1949 might have produced their first chick in 1950, but this would not have matured until the winter of 1958, by which time there could have been seven other immature birds, one of which would mature in successive years until 1965. However, without knowing the sex of these birds, it is not possible to predict future events.

It is possible that the three males and three females introduced into Hastings Caves in 1945 could have commenced to breed in that year and, on this assumption, it is possible that, at the onset of the 1953 breeding season, there could have been six mature birds from the original stock and twenty-four immature birds in the vicinity of Hastings Caves. Three of the twenty-four young birds would have matured each year, commencing in 1953; but the outcome is uncertain. Clearly, if four 'pairs', instead of three, had been introduced in 1945, there could have been four young birds maturing each year after 1953, which would have greatly facilitated pairing for breeding purposes.

Because there are so many unknown factors to be taken into account, this analysis should be accepted with due caution; but it does indicate what is involved, even under the most favourable conditions, in the establishment of a viable lyrebird population. It will be recognized, of course, that, disregarding for the moment the possibility of a polygamous relationship, two pairs of breeding lyrebirds are required in order to produce one pair of breeders at the end of eight years, and the latter will produce only one chick each year.

It is not known how many lyrebirds there are at present in Tasmania, but it can be seen that numerous stable colonies now exist in many different areas far removed from where the birds were first released in both Mount Field National Park and Hastings Caves Scenic Reserve. This suggests that the birds suffer much less from predation in Tasmania than on the mainland and it is possible that they mature perhaps a year earlier in their new home than in their old one. The climate, although severe, does not appear to have affected them adversely.

# The song of the lyrebird in Tasmania

In their natural home, the Victorian lyrebirds had been accustomed to hearing the songs of numerous birds which shared the forest with them, but in Tasmania they would not hear the hearty laughter of the kookaburra; no longer would they hear the raucous calls of the red wattlebird; but, in its stead, the lugubrious notes of the yellow wattlebird; gone would be the chuckles and resounding 'Weeeeee-ACK' of the whipbird; in place of the chatter and bell-like notes of the crimson rosella, the strange and different notes of the green rosella would be heard; the loud call of the pied currawong would give way to the rather plaintive notes of the black currawong. The Victorians might well have felt a little bewildered; but, as if to compensate them for their losses, there would be a warm welcome from the crescent honeyeater and the loud screeching of the black cockatoo, and they might have noticed the resemblance between the notes of the black-headed honeyeater and those of the familiar brown-headed honey-eater; but the grey thrush would have seemed almost like a stranger, so different are the songs of the Tasmanian thrush and its Victorian cousin. The new arrivals might have wondered at the 'chop chop' and 'rolling' notes of the yellow-throated honeyeater, but surely the familiar notes of the golden whistler and the olive whistler would have fallen pleasantly on their ears. But, most importantly, in Tasmania they would not have heard the songs of their own kind, which play such an important part in the development of the lyrebird's own song on the mainland. It is possible, of course, that the newcomers were so busy searching for food that the changes in forest voices did not bother them.

It seems fair to assume that initially the Victorian lyrebirds, when they sang, would have used the language to which they were accustomed; but, gradually, through being exposed to the calls of the birds of their new country, they would incorporate them in their repertoires. Presumably, for some time, the new calls would have mingled with the old; but, eventually, the new calls would predominate and eventually supersede the old ones. The only imitated call which I was able to recognize in the mimicry of the Tasmanian lyrebirds which resembled the notes of a bird not found in Tasmania was the whipbird. The two whistlers and the golden-bronze cuckoo are of course common to both states, so the inclusion of the notes of these birds in the repertoires of the Tasmanian lyrebirds was predictable. The same applies to the English blackbird, the grey thrush and the black cockatoo.

This may throw some light on the manner in which the lyrebird learns his repertoire, that is, by being regularly exposed to the calls of his models. The ease with such calls can be learned by a proficient songster is well illustrated by the manner in which the Hastings Caves lyrebird incorporated my own whistling in his mimicry. When birds are younger, they practise assiduously, like students of the piano learning their scales and, initially, often perform poorly.

Gordon Rowe, in February 1944, heard a mature male lyrebird, in a gully some 32 km from where the birds had first been released, imitating the kookaburra, but I did not hear any such imitations in 1980 and, although there were a few kookaburras (introduced from the mainland) to be seen near the camping ground in Mount Field National Park, during our visit, these birds did not indulge in their distinctive hearty song. It is possible that the mimicry noted by Gordon Rowe was a fading relic of earlier days, which has since been eliminated from the mimicry of the Tasmanian lyrebird.

The imitatory calls identified in the song of the lyrebirds during our visit to Tasmania were as follows:

Black currawong (*Strepera fuliginosa* Gould)
Black-faced cuckoo shrike (*Coracina novae-hollandiae* Gmelin)
Crescent honeyeater (*Phylidonyris pyrrhoptera* Latham)
Yellow-throated honeyeater (*Meliphaga flavicollis* Vieillot)
Black-headed honeyeater (*Melithreptus affinis* Lesson)
English blackbird (*Turdus merula* Linnaeus)
Eastern whipbird (*Psophodes olivaceous*)
Green rosella (*Platycercus caledonicus* Gmelin)
Grey thrush (*Colluricincla harmonica* Latham)
Grey fantail (*Rhipidura flabellifera* Gmelin)
Golden whistler (*Pachycephala pectoralis* Latham)
Olive whistler (*Pachycephela olivaceous* Vigors and Horsfield)
Scrub wren (probably *Sericornis humilis* Gould)
Yellow-tailed black cockatoo (*Calyptorhynchus funereus* Shaw)
Yellow wattlebird (*Anthochaera paradoxa* Daudin)
Golden bronze cuckoo (*Lamprococcyx plagosus* Latham)

During our visit, we heard the usual calls such as 'clonk clonk', 'scissors grinding', 'hissing', 'scolding' and the alarm call, but the 'blick blick blick…' note, so often heard in Victoria, as a prelude to a display, was not heard. This was surprising; perhaps we did not listen long enough, or perhaps we were too early in the season. What we took to be the territorial call was heard less frequently than on the mainland and was not the dominant note of mainland territorial calls. This may be because the need for a territorial call is less pronounced in Tasmania, possibly because of the wide dispersal of the birds which has occurred during the past fifty years. It would be interesting to see whether this situation changes as the lyrebird population grows.

# Dispersal of lyrebirds in Tasmania

From the small number of lyrebirds introduced into Tasmania numerous widely dispersed colonies have become established. They have now colonised the foothills of almost the entire Mount Field Range and have spread into the Tyenna Valley, even into the Weld Valley. From the liberation area near

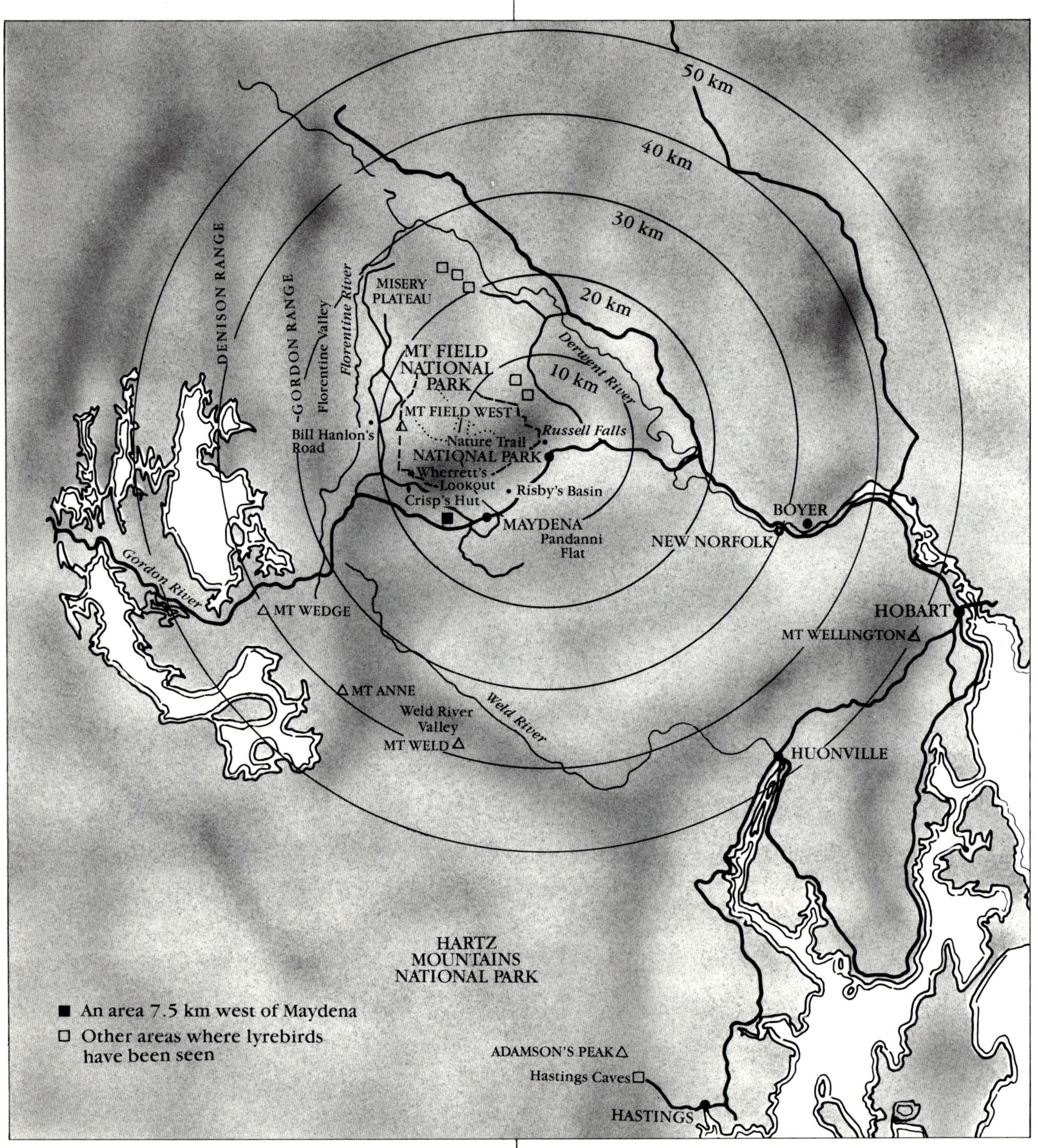


**Map 3** *Map showing where lyrebirds were released in Tasmania and areas where they have become established.*

Russell Falls, they have reached the Maydena district (12 km south-west), Bill Hanlon's Road (20 km west) and the Misery Plateau (30 km north-west), and are found through most of the Florentine Valley. Information on the dispersal of lyrebirds from the Hastings Caves colony is less readily available; but, during our visit in 1980, we heard and found lyrebirds in several localities up to five kilometres from the Caves.

Localities where lyrebirds have been seen are shown in Map 3, based on information kindly provided by the Tasmanian National Parks and Wildlife Service and Australian Newsprint Mills at Boyer.

One is tempted to ask why the lyrebirds have dispersed so widely in Tasmania. Although I am not aware of any detailed study on the ecology of the lyrebird, there is evidence from scattered sources on this subject which leads me to the conclusion that the lyrebirds are seeking better living conditions, especially food. Let me explain.

Sixty years ago, Sherbrooke Forest had a strong lyrebird population. Tregellas informs us that, in 1921, in the vicinity of the Falls, it was possible to see groups of fifteen to seventeen lyrebirds going to roost at eventide. In the 1950s, one of my delights was to watch a dozen or more lyrebirds going to their roosts in the forest near the 'firebreak', as Miss Yolande Newton, a local resident, had done thirty years earlier. One Saturday afternoon in July 1956 I counted fourteen mature male lyrebirds which paid a brief visit to Spotty's territory, filling the forest with their resounding melody. Now, that area, along with the Broadcast Area, and the forest flanking Wattle Walk, and many other areas where once the lyrebird was heard, are silent. How has this dramatic change come about?

Fifty years ago, and for many years thereafter, these conditions prevailed in Sherbrooke Forest, which became the Mecca of lyrebird seekers. The area known as the firebreak was cleared of understorey species (olearias, pomaderris, tree-ferns, etc.) and managed as a bracken area. This is the type of vegetation preferred by males for their

mounds, although mounds are also constructed among shield ferns and sword grass, in situations where there is no bracken. Nests were regularly found in the forest adjoining the firebreak and the Broadcast Area, and in the deep gullies.

Originally, the firebreak was cut and burned every two or three years, thus ensuring that there would be a good cover of bracken for the ensuing winter, but this practice ceased in the early 1960s and the firebreak has now become a jungle with a predominance of non-bracken species. There are no mounds. The Broadcast Area and the area above the Wattle Walk are likewise overgrown.

Over the years there has been a marked increase in the number of private dwellings along the periphery of the forest which now boasts green firebreaks where bracken once flourished. Timothy often displayed on mounds in the bracken between the forest and Sherbrooke Lodge Road, seemingly oblivious to passing cars.

There has undoubtedly been increased pressure on the forest fauna from domestic animals from private dwellings, and from the increasing numbers of visitors who use the picnic facilities provided by the government; but, in my opinion, these are not the major factors contributing to the decline of the lyrebird population in Sherbrooke. Even the fox, though a serious menace, is not *the* major factor. Why, then, have the lyrebirds gone?

Over the past decade or more there has been a great reduction in the area of the forest floor which is accessible to the lyrebirds, due to the ever-increasing spread of violets, ivy, wandering Jew and blackberries (all introduced species) and there has been a proliferation of wire grass, sword grass and shield ferns (native species). Areas which formerly provided food for the lyrebirds are no longer available to them. But above all, there is the problem which pervades almost the entire forest…'almost', because there are still a few areas open to the birds. This is the effect of the natural forest litter — sticks, leaves, bark, branches, fallen trees — which has been accumulating in many parts of the forest for such a long time that it now constitutes a formidable barrier between the lyrebird and its food. It is hardly necessary to say that when birds are unable to obtain a regular supply of food to enable them efficiently to perform their essential functions, they will either move to another area in search of food, or die.

There is abundant evidence to support this view. Thus, Howe (1927) reported that the lyrebirds which had escaped from the areas ravaged by bushfires, in the Warburton district in 1925, returned to their old haunts and subsequently bred there in great numbers. I do not think that they returned for sentimental reasons, but because it was so easy for them to win their food from the open ground. In recent years, I have been greatly impressed by the high lyrebird populations of the forests in the Toolangi district, near Healesville. In the several areas visited, I have heard numerous singing males on both slopes above the gullies and, at times, the volume of lyrebird song is quite remarkable. I think that it is highly significant that these slopes show evidence of recent burning for forest management purposes, and the ground cover was minimal over considerable areas, while there is still sufficient bracken cover and low scrub to provide suitable habitat for the male birds to build their mounds, which they do.

In other areas, for example, the slopes above Freestone Creek, along the old Dargo Road, in central Gippsland, we found numerous mounds in the bracken and dusty miller (*Spyridium parvifolium*) and the whole area was rich in lyrebird song. This area had been burnt in a bushfire about a year before our first visit, and the lyrebirds had easy access to their food.

During our visit to Tasmania in 1980, we investigated the matter of food supplies for lyrebirds by digging in various places frequented by the birds, but found that, even where the birds were present in reasonable numbers (the beech forest, Pandanni Flat and Hastings Caves) it was difficult to find good supplies of lyrebird food. Obviously, the lyrebirds do better than we did, but it is not easy for them. However, in the regenerating areas in the Florentine Valley, where Australian Newsprint Mills had been engaged in logging operations for some forty years, the forest litter was much less dense and one merely had to scrape away a few leaves or sticks to find lyrebird food (worms and hoppers). An essential part of the company's operations is the burning of the waste forest products resulting from the logging activities, resulting in a clear forest floor similar to those observed in the resource forests in the Toolangi district in Victoria, providing the lyrebirds with ready access to their food. These observations appear to find confirmation in those of Richard Loyn (1980) in Gippsland.

This evidence from diverse sources, although circumstantial, supports my view, I think, that the Tasmanian lyrebirds have moved away from their first homes in search of better conditions.

# 11 | Lyrebirds in Queensland

## Albert's lyrebird (*Menura alberti* Bonaparte)

It was not until about fifty years after the superb lyrebird had become known to science that the world at large learned that there was a second species of lyrebird. In 1849 a naturalist named Frederick Strange sent John Gould specimens which he had shot in the dense brushes of the Richmond River district in north-eastern New South Wales. Shortly after this, Gould received from Dr George Bennett a fine specimen belonging to the Sydney Museum. Gould named the new bird Prince Albert's lyrebird, or *Menura alberti*, in honour of Queen Victoria's consort, and prepared a paper for the Linnean Society, in 1850. However, before this was published, the French naturalist, Prince Bonaparte, presumably under the impression that Gould's paper had already been published, issued a description of the bird and to him therefore belongs the official honour of having named the new species of lyrebird.

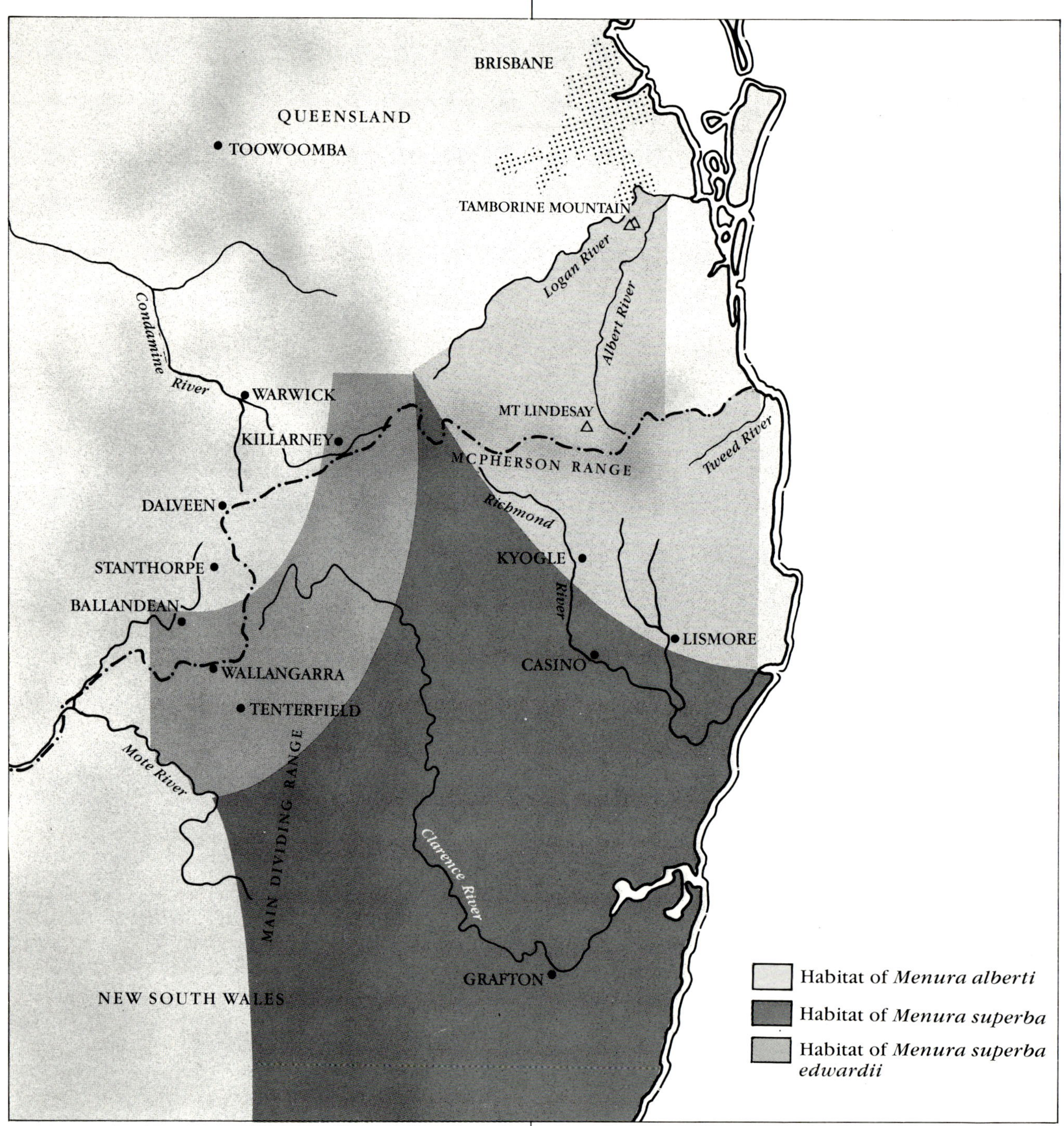


**Map** 4 *Distribution of Albert and superb lyrebirds in south-east New South Wales and south-east Queensland. From Spencer Roberts (1922), by courtesy of the Royal Australian Ornithologists Union.*

However, Strange was not the first to collect specimens of the new bird. Credit for this belongs to Augustus A. Leycester who, in 1841, established a cattle station, Tunstall, in the northern branches of the Richmond River district. Leycester shot the first specimen in the winter of 1844, but was unaware that it was a new species until 1847, when Dr Stephenson, who was familiar with the superb lyrebird, saw Leycester's specimen and realized that it was a new species. Apparently, Strange obtained his specimens in the cedar brushes which skirt Turanga Creek, Richmond River. Strange searched for a nest and egg; but, because he looked at the wrong time, without success. He thought that lyrebirds, like other birds, breed in the spring.

In the course of an adventure-packed journey into the mountains overlooking the Richmond and Tweed rivers, which began on 20 April 1859 and extended over some six weeks, Leycester collected specimens of both sexes of the new bird, and an egg, which he duly sent to John Gould, along with notes on the species. These were quoted by Gould in his description of *Menura alberti* and, because nothing of any substance had been added in the interim, Gregory Mathews, in his *Birds of Australia* (Volume VII), published in 1919, repeated Gould's account. This reflects the difficulties associated with fieldwork on this subject.

## Range and habitat

Leycester set the limits of the range of Albert's lyrebird at about 60 x 80 miles (97 x 129 km), as indicated approximately in Map 4 (taken from Spencer Roberts, 1922). However, Chisholm (1957) advances persuasive arguments, based on entries in the diary of S.W. Jackson, in support of the view that about seventy-five years ago Albert's lyrebird was also present in the Blackall Range (26/152), about 145 km north of Brisbane; but more recent studies indicate that the species no longer exists there. Chisholm (1960) represented the distribution of Albert's lyrebird and that of the superb species in its northern limits in a map prepared by Josephine Mayo, which is here reproduced as Map 5. Sadly, however, the areas occupied by the lyrebirds in these parts (as in others) have been greatly reduced by the clearing of land for farming and other purposes, and the range is now more restricted. I am indebted to Mr H.S. Curtis for the following account of the present range and habitat of Albert's lyrebird.

The range extends from the Mistake Range (27/152) south along the Great Dividing Range to the vicinity of Killarney (28/152), east along the MacPherson Range to the Springbrook Plateau (28/153) and south to Uralba (28/153) in New South Wales. However, the Uralba population stands in isolation, as does that of Tamborine Mountain (28/153) in Queensland. Fortunately, Lamington National Park and other reserves provide a permanent home for the fragmented communities which still exist and visitors to O'Reillys' Guest House and Binna Burra Lodge are assured of hearing (and possibly seeing) Albert's lyrebird in the neighbouring forests.

Albert's lyrebird lives in sub-tropical and temperate rainforest or mixed eucalypt and rainforest with adequate ground cover, sometimes the introduced lantana.

## Description

Albert's lyrebird differs considerably from the superb lyrebird. The body and wings of *M. alberti* are conspicuously more reddish-brown in colour than those of the superb lyrebird; the throat and under-tail coverts are rusty-brown; and both sexes of Albert's lyrebird have a more pronounced dark crest than the superb lyrebird. The lyrates of Albert's lyrebird are much shorter and wider, and less asymmetric than those of the other; they are also much darker (leaden-grey) on the underside and lack the orange-chestnut bars and windows of the superb lyrebird; they are also much more curved (concave to the long axis) and lack the elongated 'S' form of those of the superb lyrebird. The medians of Albert's lyrebird have much longer barbs along the greater part of the shaft than those of the superb

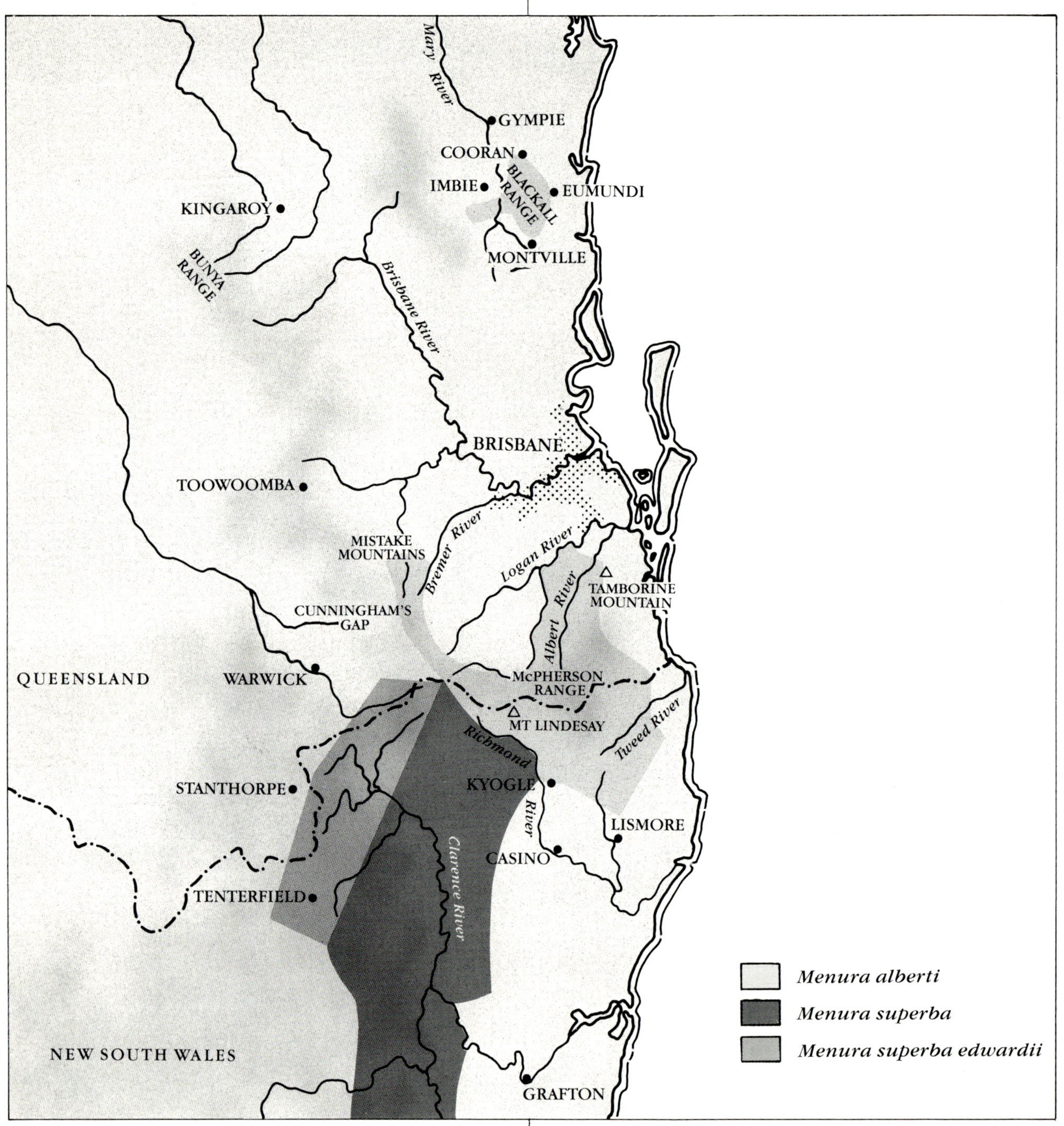


**Map 5** *Distribution of lyrebirds in north-east New South Wales and south-east Queensland, prepared by Josephine Mayo. From A.H. Chisholm (1957), by courtesy of the RAOU.*

The Albert lyrebird displaying on a platform of interlacing vines. (Photo: H.S. Curtis)

lyrebird and the filamentaries of *M. alberti* are much shorter than those of the other. Indeed, in this respect, the filamentaries of Albert's lyrebird resemble the 'first-generation' filamentaries of the superb lyrebird. Further, the curvature of the filamentaries of Albert's lyrebird, concave upwards, is much greater than it is in the case of the superb lyrebird, the medians likewise being much more curved than those of the southern bird.

The following table summarizes briefly some of the principal features of the two species of lyrebird. In body size and leg length both species are similar. The data for *M. edwardi* (page 135) have been included for ease of reference.

## General behaviour

Leycester remarked on the territorial behaviour of Albert's lyrebird, adding that he never saw more than a pair together and, from their singing, he judged that the males never approached one another closer than about 400 yards (365 m). If, as Leycester claims, Albert's lyrebird has a louder voice than the superb lyrebird of those parts, and if the forests are less dense than they are in many parts of the range of the superb lyrebird, the voice of Albert's lyrebird might carry further than does that of the other species, causing the Albert lyrebirds to keep their distance. However, this needs further study; it contrasts strongly with conditions in the Toolangi district (about 70 km north-east of Melbourne) where, during the winter, superb lyrebirds loudly proclaim themselves along the slopes at fairly regular intervals of about 100–150 m.

Detailed information on the food of Albert's lyrebird does not appear to be available. Leycester observed that 'they live entirely upon small insects, principally beetles', while Chisholm (1957) indicated that snails form an important part of the diet, at least in some parts of the range. No doubt the food eaten reflects the nature of the environment, but it is surprising that the worm is not mentioned in the meagre information available.

Leycester made some interesting observations on the development of the young birds, stating that 'the young cocks do not sing until they get their full tails which [he fancied] is not until the fourth year, having shot them in four different stages'. This differs greatly from the behaviour of the superb lyrebird and needs confirmation. In the light of modern knowledge of birds' songs, it seems improbable that such a renowned songster as Albert's lyrebird would delay its singing until the fourth year; nor does it seem likely that the male bird would have acquired a mature or full tail at this stage (see chapter 8).

## Breeding behaviour

The remoteness and rugged nature of the home of Albert's lyrebird have resulted in there being much less information available on this species than is the case for the superb lyrebird. The nesting behaviour and nest design appear to be generally similar for both species. The female Albert lyrebird does not begin daylight incubation until four days after laying the egg (Keast, 1944). The first nest and egg of Albert's lyrebird were found by James F. Wilcox of the Clarence River district, in the scrub-lands of the Richmond River, in August 1852. The nest was built on a rocky ledge about 30 m above the Richmond River and its acquisition was a feat of some magnitude. An account of this nest, along with a repro-

## A comparison of Albert's lyrebird and the superb lyrebird (museum specimens)

|  | Albert's | | Superb | | M. edwardi | |
|---|---|---|---|---|---|---|
|  | Male (mm) | Female (mm) | Male (mm) | Female (mm) | Male (mm) | Female (mm) |
| Total length | 830 | 700 | 1000 | 800 | 935 | 845 |
| Lyrate | 380 | 270 | 650 | 300 | 625 | 350 |
| Median | 535 | 420 | 710 | 480 | 670 | 490 |
| Main rectrices (filamentaries) | 520 | 360 | 630 | 370 | 640 | 430 |
| Tarsus | 114 | 105 | 114 | 104 | 108 | 100 |

duction of Gould's plate, appeared in the *Illustrated London News* of 19 March 1853 (Jackson, 1907).

The egg was described by Jackson (1907) as being quite different from that of the superb lyrebird, being a rich purplish-grey colour, the larger end having 'a cap of dark purplish-brown' with darker splotches over the remainder of the egg. This observation was based on an egg taken by Jackson, in the Booyong Scrub above the Richmond River, on 6 October 1899. The date is interesting, because it suggests that there are late layers among the Albert species, just as there are in the superb species. (Leycester collected an egg early in June 1859.) However, the egg of *M. alberti* varies in size, colour and shape, similarly to that of the superb lyrebird. The H.L. White Collection at the National Museum, Melbourne, contains several eggs of *M. alberti* which, while mainly dark to very dark, include a very light (putty-coloured) specimen, as mentioned by Gould and Mathews. The eggs in the H.L. White Collection ranged in length from 57.7 to 62.3 mm and in maximum diameter from 40.1 to 44.0 mm, the averages (n = 6) being 59.5 x 42.2 mm respectively.

The mating of Albert's lyrebird was observed by Colin Harman, curator of the orchid garden at O'Reillys' Guest House, on 14 March 1971. Coition was preceded by prolonged singing and displaying by the male, observed by Mr Harman from a distance of six metres; but, when the female presented herself to the male, he ceased the loud singing and began a slow clucking noise which became faster as mating proceeded. After mating, the two birds departed together, the male clucking softly alongside the female. Within a few minutes, the clucking was replaced by the territorial call.

This is a remarkable record, especially in regard to the date; because, if the laying of a fertile egg followed, as usual, the chick would have left the nest by about the second half of June, which is the time when most superb lyrebirds are engaged in mating activities. It would mean, also, that the two lyrebirds had come into breeding condition about three months before their southern cousins.

## Courtship display

My own experience with Albert's lyrebird is limited to several sightings in Lamington National Park, in the vicinity of Binna Burra Lodge and O'Reillys' Guest House, but I have often heard the song from the depths of the forest.

One day, towards the end of May 1951, I was walking from O'Reilly's to Echo Point, a notable feature on the border between New South Wales and Queensland, when I was drawn off the track by the song of an Albert lyrebird. At first I could not see the bird; but, after some careful stalking, I was rewarded with a view of a shimmering haze of feathers from the midst of which came a stream of notes with which I was then unfamiliar. In my enthusiasm, I pressed closer, but suddenly found myself caught in a lawyer vine and, in my efforts to extricate myself, caused the bird to dissolve into the green light. Eventually I escaped and decided to inspect the mound on which the bird had been displaying, but was mystified when I discovered that there was no mound. Many years later, I learned that Keast had had a similar disappointment in 1944.

The mystery remained until Syd Curtis (1972) demonstrated that Albert's lyrebird does not use a mound, but stands on a platform of several interlacing vines. Like his southern cousin, he also displays on rocks, stumps, logs and, sometimes, on open ground; but the platform of vines takes the place of the mound of the superb lyrebird. Whether this interesting adaptation is due to the nature of the terrain or the ready availability of vines is not immediately apparent, but it is a great labour-saving device and may be because the much smaller tail of Albert's lyrebird makes it possible for him to display in a more confined space than the superb lyrebird requires.

It was this detailed knowledge of the habits of Albert's lyrebird which enabled Syd Curtis, with infinite patience, to take the first photographs of this shy bird displaying, by setting up his camera near a platform and operating the camera by remote control. This was as recent as 1968.

Beautiful as it is, simply because of the smaller tail and the lack of ornamentation in the side feathers (lyrates), the display of Albert's lyrebird is less spectacular than that of the superb lyrebird. Both species have the ability to manipulate the tail feathers into a variety of intermediate positions as well as the full-face arrangement. Curtis observes that, in the course of the display, without moving his feet, Albert's lyrebird, like the superb, sways his body from side to side, and both species move their heads from time to time, possibly to enable them to keep a closer watch on their surroundings. Another interesting feature of the display of Albert is the 'stepping from one foot to the other' in a manner similar to that adopted by the superb lyrebird. In the latter case, his rather boisterous movement is usually accompanied by the 'clonk clonk' call and its variants; but, in the tapes of the song of Albert's lyrebird which I have studied, these items appear conspicuous by their absence. The superb lyrebird, also, in the course of his display (sometimes on a mound and sometimes on the floor of the open forest) performs what has been called, for want of a better name, the 'dance'. In this movement, the bird moves his whole body and feet in the form of a rough square — to the right, to the rear, to the left, forward (back to the original position). It is not clear whether Albert's lyrebird performs a similar 'dance'.

The story concerning the display of Albert's lyrebird which perhaps surpasses all others is that told by W. Gordon Lawler (1982), whose almost fanatical enthusiasm for wild-life photography brought him rich rewards. After several years of preparatory work, in July 1979 he located a lyrebird's display platform and built a hide (or blind) near the platform. But this was not the sort of timber-framed structure on which I drape a small square-topped tent in which I secrete myself when I want to photograph the rainbow bird or the grey thrush, etc. No — this was a pit, about 100 cm x 60 cm x 60 cm, which he dug in the ground about four metres from the spot where the lyrebird was expected to perform. With the excavated earth piled up around the perimeter of the pit, he gained a little extra height (or depth); the pit was then covered by a large log and forest litter. Somehow, he managed to crawl into the hide, set up his tripod, camera and flash-gun, and focus on the target area. In the dim light of a cold wintry dawn, the lyrebird arrived and began to display *on top of the hide*, but then it graciously moved to the platform, where it continued its display while Lawler exposed two spools of colour film. The results were spectacularly superb and would surely be the envy of all other nature photographers.

It seems remarkable that this lyrebird, renowned for his timidity, which occasioned the use of such an unusual hide, should have been so tolerant towards the flash-gun.

## Song of Albert's lyrebird

The song of Albert's lyrebird is loud and striking but opinions vary as to how it compares in quality with that of the superb lyrebird. Leycester went so far as to say that this lyrebird was 'far superior' to the superb lyrebird in its 'powers of mocking and imitating the cries and songs' of other birds' songs, and he was perhaps the first to remark that 'its own peculiar cry or song is also different, being of a much louder and fuller tone'. He wrote,

'I once listened to one of these birds which had taken up its quarters within 200 yards of a sawyer's hut — he had made himself familiar with all the noises of the sawyer's homestead — the crowing of the cocks, the cackling of the hens, the barking of the dogs, and even the screechings of the sharpening or filing of the saw. '

Presumably Leycester was comparing *M. alberti* with the superb lyrebird of north-eastern New South Wales; some people who are more familiar with the songs of the superb lyrebird of other parts may have reservations about accepting Leycester's verdict in favour of *M. alberti*. Personally, I think that beauty of song resides in the ear of the listener, and I find them both exciting and interesting. But I must agree with Leycester that Albert's lyrebird has a powerful voice and a loud and distinctive territorial call.

Albert's lyrebird has long been an active interest of Mr H.S. Curtis who, in 1970, in collaboration with Norman Robinson, began a comprehensive study of the song of this bird, and I am grateful for the opportunity to analyse some of his excellent recordings.

The song of Albert's lyrebird, like that of his southern counterpart, consists of a territorial call and a range of mimicry of the calls of other birds. The territorial call comprises a much greater part of the total song than does that of his neighbour, the superb lyrebird of Girraween, but many of the specific items in the song of the southern species appear to be absent from the song of Albert's lyrebird. These include the 'clonk clonk' and the 'clonk clonk clickety clickety clickety click' items, and the 'blick blick…' and 'clack clack…' calls. In the mimicry of Albert's lyrebird, the song of the bowerbird, which itself is complex, occupies a prominent place, and the somewhat simpler call of the spine-tailed logrunner is also frequently heard. Both of these birds are active in the domain of Albert's lyrebird.

On the other hand, the grey thrush has only a minor role in the mimicry of *M. alberti*, and the pied currawong and the whipbird, which are fairly common in its range, are seldom, if ever, heard in the Albert's winter song — at least, not in the tapes which I studied. These were made in July 1983, that is, at about the peak of the breeding season. Robinson (1974) has pointed out that the breeding seasons of Albert's lyrebird and the whipbird of south-eastern Queensland coincide and has argued that the inclusion of the whipbird's notes in this lyrebird's mimicry would be disgenic to that species.

In the case of the superb lyrebird and the whipbird of Sherbrooke Forest and other parts of the range, the latter species breeds later than the lyrebird; hence the inclusion of the whipbird's notes in the mimicry of the superb lyrebirds of those parts does not entail this penalty.

The following items were identified in the song of Albert's lyrebird:

Territorial call
Satin bowerbird
Spine-tailed logrunner
Crimson rosella
Kookaburra
Snapping of the beak of the kookaburra
Wings of pied currawong in flight
King parrot
Grey thrush
Paradise riflebird
Yellow robin
Lewin honeyeater
White-eared honeyeater
Catbird
Thornbill (?)

# Prince Edward's lyrebird (*Menura edwardi*)

When thinking of the home of the lyrebird generally, one envisages densely wooded hills and slopes, rocky watercourses, fern gullies and rich soils which provide an abundance of food for the birds. One thinks of places like Sherbrooke Forest, the Dandenong Ranges, the mountain gullies of the Healesville and Warburton districts, the Gippsland forests and the eastern highlands of New South Wales extending up to the Queensland border. But here, curiously, a different situation is met. Standing in geographical isolation, about 160 km inland, the 'Granite Belt' runs in a south-westerly direction from the junction of the Main Dividing Range and the MacPherson Range (which runs from west to east, down to the coast) back to the Main Dividing Range. At some time in the historic past, this area may well have been taller, clothed in dense forests; but, over the ages, erosion has eaten into this land mass, leaving groups of granite boulders distributed within a fairly open forest. The vegetation consists of a variety of eucalypts (including stringy-barks which are, however, of lower stature than in other parts) acacias,

*A fern gulley in Sherbrooke Forest, once renowned as the home of the superb lyrebird.*

*Callitris* spp. and casuarinas, with an understorey of banksias, tea-trees and heathy plants — not, one would have thought, the lyrebird's first choice of a home. Yet, here the lyrebird dwells, along with the echidna (spiny anteater), platypus, wombat and a variety of birds. The sandy soil supports about 400 species of wildflowers. This area has been incorporated in the Giraween National Park.

Some 20–30 km to the west lies the Sundown National Park, spanning the Severn Valley and encompassing some deep gorges; here, too, the superb lyrebird is found, but the lyrebirds of this colony appear to be less well known than those of Girraween.

## Discovery

The lyrebird of the Granite Belt was brought to public notice through the combined efforts of several people. In 1915 Dr Spencer Roberts went to live in the Stanthorpe district, in south-eastern Queensland, in the hope of improving his wife's health. He soon became aware of the existence of lyrebirds nearby, because of the numerous tail feathers which adorned the homes of his patients, but did not make the acquaintance of the bird until 1919 when, under permit, he collected two eggs destined for the H.L. White Collection. The first to be received was described by H.L. White as the type for *Menura novaehollandiae edwardi*. In 1921, just before the onset of the breeding season, Roberts collected two males and a female, one of the former being sent to the H.L. White Collection and subsequently described by A.H. Chisholm (1921) as the type for the new species (see below). The remaining two specimens were later added to the H.L. White Collection.

While living in Stanthorpe, Roberts became acquainted with Alec Gemmell, whose property adjoined a colony of lyrebirds. In 1920 Gemmell called on A.H. Chisholm to give him a report on the lyrebirds of the Granite Belt on behalf of Dr Roberts. The doctor's descriptions of this interesting lyrebird and its home galvanized Chisholm into action. Probably, too, his activity was catalysed, if some-

what belatedly, by a report in 1919 from a local lad apprising him of the existence of the colony of lyrebirds in the Stanthorpe district.

In August 1920 Chisholm spent three days among the lyrebirds and rocks of the granite belt and later (1921) described the bird as a new species of lyrebird which he named Prince Edward's lyrebird or *Menura edwardi*, in honour of Edward, Prince of Wales, who visited the district in July 1920.

## Designation

There are interesting differences, some quite striking, between the lyrebird described by Roberts and Chisholm and those of more southern parts. The lyrates of the northern bird are much less curved than those of the south; the black tips are smaller and the pigmented areas between the windows in the lyrates of the northern bird are lemon-yellow in contrast with the yellow-orange-chestnut colours of the lyrates of the better-known superb lyrebird. The distal ends of the barbs of the lyrates of both sexes of Roberts's bird are black, to the extent of about 6 mm, a feature which is conspicuously absent from the lyrates of the female superb lyrebird of more southern parts, as noted also by Campbell (1922). It would be interesting to examine the lyrates of females from the more northerly parts of the range, in such districts as Dorrigo and New England National Park, to see whether such differences exist there also.

However, despite these differences, there appears now to be a general concensus among ornithologists that, to quote A.H. Chisholm, *M. edwardi* is really 'a southern lyrebird in a northern setting', just as the former *M. victoriae* is nowadays regarded as the southern form of *Menura novaehollandiae*.

In what follows, the designation 'superb lyrebird' will be used, but it must be mentioned that my knowledge of the subject is based on reports published by Roberts in 1922 and 1926 and the examination of specimens in the National Museum (Melbourne) and the Australian Museum (Sydney).

# Range

Roberts reported that the range of the lyrebird in the Granite Belt was about 30 x 30 km² (although it was originally greater, before farming developments reduced it); but in the light of changes which have occurred during the past sixty years, some revision may be necessary. However, Map 4, which is taken from Roberts's original paper, indicates the general relationships between the ranges of the Albert lyrebird, the superb lyrebird (*M. novae-hollandiae*) and the subject of Robert's report.

I do not know whether the superb lyrebird of the Sundown National Park is identical with that of Girraween, but it would be surprising if the lyrebirds from these two areas did not differ in some ways. The superb lyrebirds from different districts within the range, and even within a given area, show remarkable variations in respect of body colour, size and pigmentation of lyrate feathers, dimensions and shape of median tail feathers and the filamentary feathers. There are also striking differences between the territorial and other specific calls and between the dialects of their songs. Indeed, it would be fair to say that the 'type' superb lyrebird is not precisely duplicated in any other part of the range of the species. And other wide-ranging species of birds, such as the grey thrush, the pied currawong and the whipbird, show dialectal differences in their songs throughout their respective ranges.

# Behaviour

Like the familiar superb lyrebird of more southern parts, that of Girraween is said to be communal in behaviour during the non-breeding period. It would be surprising were it otherwise, because this communal living is an important part of the young lyrebirds' training. At the approach of winter, the males resume their territories in which they build their display mounds, while the females 'do battle over them'. It is not clear from Roberts's report whether the females actually fight over the males, but it is known that some female lyrebirds in Sherbrooke Forest behave aggressively towards other females during the breeding season, although they are mostly at peace with one another.

It is exceptional in the Girraween area for nests to be built at ground level, the preferred site being a niche or ledge on a granite rock, and most nests lack the camouflage which is such a feature of those of the southern birds. Nests may be up to about eight metres above the ground, averaging 2.5 metres. Roberts says that the nest takes about two months to build and that it is based on an old nest site which has a secure hold on the granite. The single egg is laid in July; eggs vary in colour from very light to dark and appear to be somewhat smaller than the average egg of the superb lyrebird in other areas.

As is customary, the female provides a bed of feathers from her flanks before laying the egg and leaves the nest unattended for several days before commencing incubation. Roberts says that incubation extends over four weeks, possibly a little longer, and the chick (which resembles that of the superb lyrebird of other parts) remains in the nest for at least seven weeks before leaving and 'vanishing from view'. Especially because of the lower temperatures prevailing in the Granite Belt, I think that the shorter incubation period proposed by Roberts requires confirmation.

There are other aspects of the behaviour of these birds which would repay closer study. Roberts says that the lyrebirds of the Granite Belt do not actually scratch, but carefully pick up the litter and place it aside to expose the beetles and snails which appear to be the chief items of food. The proclivity of the superb lyrebird of other parts to scratching is well known, and Sherbrooke Forest contains many terraces resulting from the lyrebirds' habit of scratching downhill. Kitson (1905) stated that, in Gippsland, certain tracks had been rendered unusable because they were littered with logs which had rolled downhill as a result of the lyrebirds' having scratched the soil away from the lower side. Moreover, Albert's lyrebird is a renowned scratcher and a great opportunist. In a letter to Mr Edward Vidler (1933), Mrs Margaret Hardcastle stated that, when the rabbit-proof fence was being constructed in the Condamine Scrub district, Albert's lyrebird often filled in the trenches which had been dug in which to bury the bottom of the wire-netting — the sight of all that freshly dug earth was too much of a temptation for them. It has not taken the lyrebirds long to find the tracks which I have cut through the scrub in order to facilitate my approach when recording the lyrebirds' songs in different parts of Victoria, and full advantage has been taken of the easy access which the tracks provide to the lyrebirds' food supplies. So this dainty habit of the lyrebirds described by Roberts must be seen as something very unusual.

## Song

The song of the lyrebird of Girraween National Park is remarkable for both its quality and its content. I have not yet met this bird 'in person', and what follows is based on my analysis of a recording made by H.S. Curtis who made it available to me for this purpose.

Despite the apparent isolation of the superb lyrebird of Girraween, the general structure of this bird's song resembles that of the superb lyrebird of southern parts, but there are some striking differences, also. The territorial call is a delight to human ears, consisting as it does of a succession of trills, produced with consummate artistry. In structure it reminds me of the territorial call of the superb lyrebird of Tidbinbilla; both consist of a staccato note which is repeated often before falling away and finally ceasing; but, to my ears, the call of the northern bird is softer and is rendered in a higher key than the other.

As with the superb lyrebird of southern parts, the song of the Girraween lyrebird consists of a number of specific lyrebird items blended randomly with some delightful mimicry, but the specific items show more pronounced divergence from their southern counterparts than do the items of mimicry. One recognizes the 'clonk clonk' call and its variant 'clonk clonk clickety clickety clickety click', but the northern and southern forms of this item are different, and what is presumably the 'scissors grinding' item shows even greater divergence from its southern equivalent. There are certain items in the repertoire of the Girraween bird which appear to have no counterpart in the song of the southern superb lyrebird.

The mimicry of this lyrebird is remarkably clear and varied. The song of the grey thrush, with its many variants, occupies a prominent place in his mimicry, and the notes of the white-eared honeyeater are imitated with precision. The satin bowerbird, with its own extensive repertoire, is well imitated and the crimson rosella is also one of this lyrebird's favourites, although its several notes differ somewhat from those of the crimson rosella so faithfully imitated by the lyrebirds of Sherbrooke.

One item consists of a series of notes reminiscent of the whipbird's notes; but, as the whipbird is stated not to occur in this area, the identity of this item is somewhat clouded. It could be a relic of bygone days (as in the case of the lyrebirds in Tasmania), but the matter needs further study. However, a truly remarkable item of mimicry was the creaking noise made by two trees rubbing against one another. The frequent mimicking of the kookaburra clicking his beak and tapping his beak against a tree suggests that the kookaburra and the lyrebird are fairly closely associated in Girraween National Park.

The following items of mimicry were identified in the song of the superb lyrebird of Girraween:

Grey thrush (with variants)
Satin bowerbird (various notes)
Crimson rosella (various notes)
White-eared honeyeater (several notes)
Scrub wren
Brown thornbill
Black cockatoo (several notes)
Pied currawong (with variants)
Kookaburra
Kookaburra snapping beak
Kookaburra tapping beak against a branch

Yellow robin
Golden whistler
Wattlebird
Whipbird (?)
Crescent honeyeater or yellow-faced honeyeater
Creaking of two trees rubbing together

# References

Aflalo, F.G. (1896), *A Sketch of the Natural History of Australia*, London.

Barruel, P. (1973), *Birds of the World*, Oxford University Press, New York.

Bartlett, A.D. (1867), 'Notes on the habits of the lyrebird', *Proc. Zool. Soc.,* London, 232.

Becker, L. (1857), 'On the nest, egg and young of the lyrebird', *Trans. Philos. Inst. Vict. 1,* 153–4.

Beilby, J.W. (1858), 'On the lyrebird of Gippsland', *Trans. Philos. Inst. Vict. 2,* 12–14.

Bellairs, R. (1960), 'Development of Birds', in *Biology and Comparative Physiology of Birds* (ed. A.J. Marshall), vol. 1, ch. 5.

Bennett, G. (1860), *Gatherings of a Naturalist in Australia*, London.

Bill, M.E. (1932), 'Lyrebirds and bushfires', *Vict. Nat. 49,* 24.

—— (1933*a*), 'A day with a lyrebird', *Vict. Nat. 50,* 74.

—— (1933*b*), 'A white lyrebird', *Vict. Nat. 50, 26.*

Blakers, M., Davies, S.J.J.F. and Reilly, P.N. (1984), *The Atlas of Australian Birds.* Melbourne University Press.

Bradford, S.L. (1933), 'Notes on the incubation of the lyrebird', *Emu, 33, 230.*

Broinowski, G.J. (1891), *Birds of Australia.* Charles Stuart & Co. Melbourne, Sydney, Adelaide, Brisbane, New Zealand and Tasmania.

Brunner, H., Lloyd, J. and Coman, B.J. (1975), 'Fox scat analysis in a forest park in south-eastern Australia', *Aust. Wildl. Res. 2,* 147–54.

Brush, A. H. and Wyld, J. A. (1982), 'The Molecular Organisation of Epidermal Structures', *Comp. Biochem. Physiol., 73b,* 313–25.

Campbell, A.G. (1941), 'Courtship of the lyrebird', *Emu, 40, 357–64.*

—— and Gray, A. (1942), 'Lyrebirds of Sherbrooke', *Emu, 42,* 106–11.

Campbell, A.J. (1884), 'Notes about lyrebirds', *Vict. Nat. 1,* 105–8.

—— (1900), *Nests and Eggs of Australian Birds,* Sheffield.

—— (1922), 'The lyrebird *Menura novaehollandiae* Latham, *Emu, 21,* 241.

Cayley, N.W. (1949), *The Fairy Wrens of Australia,* Sydney.

Chisholm, A.H. (1921), 'A new menura: Prince Edward's lyre-bird', *Emu, 20,* 221–3.

—— (1957), 'The Albert lyrebird — a puzzle in distribution', *Emu, 57,* 25–30.

—— (1960), *The Romance of the Lyrebird.* Sydney.

—— (1965), 'Further remarks on vocal mimicry', *Emu, 65,* 57–64.

Collins, D. (1802), *An Account of the English Colony in New South Wales,* vol. 2, 320.

Cook, L.C. (1909), 'The lyrebird at Poowong', *Emu, 8,* 220–1.

—— (1916), 'Notes on the lyrebirds of the Poowong district', *Emu, 16,* 101.

Curtis, H.S. (1972), 'The display of the Albert lyrebird', *Emu, 72,* 81–4.

Davies, T. (1802), 'Description of *Menura superba,* a bird of New South Wales', *Proc. Linn. Soc. Lond. 6,* 207–9.

Delacour, J. (1951), *Pheasants of the World,* London.

Denton, Sherman F. (1889), *Incidents of a Collector's Rambles in Australia, New Zealand and New Guinea,* Lee and Shepherd, Boston.

D'Ewes, J. (1853), *Sporting in Both Hemispheres,* London.

Dorst, Jean (1971), *The Life of Birds* (trans. I.C.J. Galbraith), Weidenfeld & Nicolson, London.

Dwight, J. (1900), 'The sequence of plumage moults in the passerine birds of New York', *Ann. N.Y. Acad. Sci. 13,* 73–345.

Eddy, R.J. and Cusack, F. (1967), 'The lyrebirds of Wandong, Victoria', *The Australian Bird Watcher, 3* (no. 1), 14–19.

Edwards, H.V. (1919), 'The nesting of lyrebirds', *Emu, 18,* 298–9.

Eisner, E. (1960), 'The relationship of hormones to the reproductive behaviour of birds', *Animal Behaviour, 8,* 155–79.

Feduccia, A. and Olsen, S.L. (1982), 'Morphological similarities between menuras and the rhinocryptidae, relict passerine birds of the southern hemisphere', *Smithsonian Contributions to Zoology,* no. 366.

Fleay, D. (1941), 'Lyrebirds adopt a foster child', *Vict. Nat. 57,* 207–8.

—— (1942), 'Now is the concert-time for lyrebirds', *The Australian,* 27 June.

—— (1952), 'Transporting lyrebirds to Tasmania', *Vict. Nat. 69,* 60–3.

Fox, H.M. and Vevers, G. (1960), *The Nature of Animal Colours,* London.

Frieden, E.H. (1976), *Chemical Endocrinology,* Academic Press, New York.

Gillan, Miss (1920), in *The Land of the Lyrebird,* Gordon & Gotch Australasia.

Glegg, W.E. (1944), 'Barred feathers in birds', *Ibis, 87,* 471–3.

Godfrey, F.P. (1905), 'Domesticated lyrebirds', *Emu, 5,* 33–4.

Gould, J. (1848), *Birds of Australia, Vol. 3* and supplement, London.

Grey, J.E. (1866), *Proc. Zool. Soc. Lond.* 167–8.

Halafoff, K.C. (1959), 'Musical analysis of the lyrebird's song', *Vict. Nat. 75,* 169–78.

—— (1959), 'The range of the lyrebird's song', *Vict. Nat. 76,* 121.

—— (1961a), 'Notes on the lyrebird's song', *Vict. Nat. 78,* 79–81.

—— (1961b), 'Writing down a lyrebird's song', *Vict. Nat. 77,* 335–8; *ibid. 77,* 359–63.

—— (1964), 'Audiospectrographic analysis of the lyrebird's song', *Vict. Nat. 80,* 304–12.

—— (1962), 'The local dialects of Gippsland lyrebirds', *Vict. Nat. 79,* 137–9.

Harrison, C.J.O. (1964), *Birds' Nests and Eggs,* Museum Press, London.

Hartshorne, C.E. (1956). 'The monotony threshold in singing birds', *Auk, 73.* 176–92.

Haydon, G.H. (1846), *Five Years' Experience in Australia Felix,* London.

Hinde, H.A. (1962), 'Temporal relationships of brood patch development in domesticated canaries', *Ibis, 104,* 90–97.

Hindwood, K.A. (1941), 'Nesting habits of the superb lyrebird', *Vict. Nat. 57,* 183–8; 194–202.

—— (1955), 'Long use of nest by lyrebird', *Emu, 55,* 257–8.

—— (1960), 'Two eggs and second eggs in lyrebirds' nests', *Australian Bird Watcher, 3,* 94–6.

Höhn, E.O. (1950), 'Physiology of the thyroid gland in birds; a review', *Ibis, 92,* 464–73.

—— (1960), 'Endocrine Glands, Thymus and Pineal Body', in *Biology and Comparative Physiology of Birds* (ed. A.J. Marshall), vol. 2, ch. 16.

—— (1966), *Hormones in Man and Animals,* The English Universities Press, London.

Howe, F.E. (1927), 'Nesting habits of the lyrebird', *Emu, 27,* 117–18.

Ireland, T. (1950), *The Birds of Paradise and Bower Birds,* Georgian House, Melbourne.

Jackson, S.W. (1907), *Catalogue of the Jacksonian Oological Collection.*

Keast, A.J. (1944), 'A winter list from the Tweed River district of New South Wales, with remarks on nomadic species', *Emu, 43,* 177–87.

Kenyon, R.F. (1972), 'Polygyny among superb lyrebirds in Sherbrooke Forest Park', *Emu, 72,* 70–6.

Kitson, A.E. (1905), 'Notes on the Victorian lyrebird (*Menura victoriae*)', *Emu, 5,* 57–67.

Lawler, W.G. (1982), 'Crisis in the rain forest', *Australian Photography, 33* (no. 2), 40–3.

LeSouef, D. (1903), 'Male lyrebird incubating', *Emu, 2,* 173.

Lewin, R. (1972), *Hormones — Chemical Communicators,* London.

Leycester, A.A. (1880), 'Adventures of an early naturalist on the Richmond', *Sydney Mail,* 10 July.

Lill, A. (1979), 'Nest inattentiveness and its influence on the development of the young in the superb lyrebird', *Condor, 87,* 225–31.

—— (1979), 'An assessment of the male parental investment and pair bonding in the polygamous superb lyrebird', *Auk, 96,* 489–98.

—— (1980), 'Reproductive success and nest predation in the superb lyrebird', *Aust. Wildl. Res. 7,* 271–80.

Lillie, F.R. and Juhn, M. (1932), 'The physiology of the development of feathers. I Growth rates and pattern in the individual feather', *Physiol. Zool. 5,* 124–84.

Littlejohns, R.T. (1926), 'Filming lyrebirds', *Emu, 25,* 271.

Lonsdale, T. (1987), 'Friends from the bush', *BOC Newsletter,* 663 (May), 46.

Mack, G. (1952), 'The Lyrebirds in Queensland', *Vict.Nat., 69,* 68–70.

Mahan, J. C. (1905), in 'Nature Notes' by D. McDonald, in the *Argus,* Melbourne, 28 July 1905.

Maroney, D. (1972), 'Plumage changes in the superb lyrebird', *Emu, 72,* 17–21.

Marshall, A.J. (1954), *Bower-birds: Their Displays and Breeding Cycles,* Clarendon Press, Oxford.

Mathews, G.M. (1919), *Birds of Australia,* vol. 7, London, 392–407.

—— 'G.J. Broinowski', *Emu, 41,* 195–9.

Newton, A. (1890). *Encyclopaedia Britannica,* vol. XV.

Nicholls, E.B. (1911), 'A trip to the Bass Valley', *Vict. Nat., 28,* 149–57.

O'Donoghue, J.G. (1914), 'Notes on the Victorian lyrebird *Menura victoriae*', *Vict. Nat. 31,* 11–20.

O'Reilly, V. (1971), 'The mating dance of the Albert lyrebird', *BOC Notes,* no. 474 (May), 5–6.

Portman, A. (1963), *Animals as Social Beings,* London.

Pratt, A. (1933), *The Lore of the Lyrebird,* Melbourne.

Rawnsley, H.C. (1863), 'The lyrebird of Australia', *Trans. Philos. Soc. Qld. 1,* 1–3.

Reilly, P.N. (1970), 'Nesting of the superb lyrebird in Sherbrooke Forest, Victoria', *Emu, 70,* 73–8.

Ridpath, M.G. and Moreau, R.E. (1966), 'The birds of Tasmania: ecology and evolution', *Ibis, 108,* 348–93.

Roberts, S. (1922), 'Prince Edward's lyrebird at home', *Emu, 21,* 242–53.

—— (1926), 'Prince Edward's lyrebird re-visited', *Emu, 26,* 105–9.

Robinson, F.N. (1974), 'The function of vocal mimicry in some avian displays', *Emu, 74,* 9–10.

—— (1975), 'Vocal mimicry and the evolution of bird song', *Emu, 75,* 23–7.

—— (1977), 'Environmental origins of the *Menurae*', *Emu, 77,* 167–8.

—— and Frith, H.J. (1982), 'The superb lyrebird *novae-hollandiae* in Tidbinbilla (ACT)', *Emu, 82,* 145–57.

Rowley, I. (1974), *Bird Life,* Collins, Sydney and London.

Serventy, D.L. (1973), 'Origin and structure of Australian bird fauna' in *Birds of Australia* (J.D. Macdonald), A.H. & A.W. Reed.

Sharland, M.S. (1931), 'The lyrebird on the air', *Emu, 31,* 231–2.

—— (1944). 'The lyrebird in Tasmania', *Emu, 44,* 64–71.

—— (1952), 'The lyrebird in Tasmania', *Vict. Nat. 69,* 58–9.

—— (1981), 'Memories of Tom Tregellas', *Australian Bird Watcher, 9*(4), 103–9.

Sibley, C.G. (1974), 'The relationships of the lyrebirds', *Emu, 74,* 65–74.

Sibley, C.G. and Ahlquist, J.E. (1985), 'The phylogeny and classification of the Australian–Papuan passerine birds', *Emu, 85,* 1–21.

Smith, L.H. (1968), *The Lyrebird,* Lansdowne Press, Melbourne.

—— (1982), 'Moulting sequences in the development of the tail plumage of the superb lyrebird *Menura novae-hollandia*', *Aust. Wildl. Res. 9,* 311–30.

Stone, A.C. (1916), 'Porosity of the lyrebird's egg', *Emu, 16,* 109.

Terres, J.K. (1980), *The Audubon Encyclopaedia of North American Birds,* Alfred A. Knopf, New York.

Thomson, D.E. (1934), 'Some adaptations for the disposal of the faeces. The hygiene of the nest in Australian birds', *Proc. Zool. Soc.,* part 4, 701–7.

Thorpe, W.H. (1961), *Bird-song: The Biology of Vocal Communication and Expression in Birds,* Cambridge University Press.

Tregellas, T. (1921), 'Further notes on the lyrebird', *Emu, 21,* 95–103.

—— (1930), 'The truth about the lyrebird', *Emu, 30,* 243–50.

Vidler, R.A. (1933), 'Notes on the lyrebird and the satin bower-bird', *Emu, 33,* 132.

Voitkevich, A.A. (1966), *The Feathers and Plumage of Birds,* London.

Wall, L.E. and Wheeler, R.W. (1966), 'Lyrebirds in Tasmania', *Emu, 66,* 123–31.

Wallace, G.J. (1963), *An Introduction to Ornithology,* Macmillan.

Ward, J.E. (1933), 'In the haunts of the lyrebird. Foxes and feral cats and egg collectors are bringing a famous Australian bird in danger of extinction', *Bull. New York. Zool. Soc., 42,* 67–79.

—— (1940), 'The passing of the lyrebird. Rapid disappearance of one of Australia's most famous birds was discovered during investigation of its incubation period', *Bull. New York Zool. Soc., 43,* 146–52.

Watson, G.E. (1963), 'The mechanism of feather replacement during natural moult', *Auk, 80,* 486–95.

Watson, I.M. (*ca.* 1950), *Silvertail — The Story of the Lyrebird,* John Sands, Sydney.

—— (1965), 'Mating of the superb lyrebird', *Emu, 65,* 129–32.

Welty, J.C. (1982), *The Life of Birds.* Saunders College Publishing.

Westerkov, K. (1956), 'Incubation temperature of the pheasant *Phasianus colchicus*', *Emu, 56,* 405–20.

Williams, F.J. (1881), 'The habits of the lyrebird', *Southern Science Record, 1,* 87–9.

Witschi, E. (1960), 'Sex and Secondary Characteristics' in *Biology and Comparative Physiology of Birds* (ed. A.J. Marshall), vol. 2, ch. 17.

Wood, H.B. (1950), 'Growth bars in feathers', *Auk, 67,* 486–91.

# General index


*Account of the English Colony of New South Wales* (1802), 4–5
Adelaide Zoo, 105–6
adoption, chick, 38
aggression, female, 41; *see also* fighting
alarm, lyrebird during, 95–7
alarm call, 95, 97, 99, 100, 121
Albert's lyrebird, 98
    breeding behaviour, 130–1
    described, 127–30
    discovery, 125–7
    first nest and egg, 131
    range and habitat, 127
albino lyrebirds, 26
Antarctic beech, 14–15, 115, 124
'arena birds', lyrebird as, 27
artists, depicting lyrebirds, 5,7
*Atrichus* (Atrichornis), 14
Audubon, J.J., 7
Australian Broadcasting Commission, 11
Australian Museum (Sydney), 135

banding, identification, 11, 19, 79
    chicks, 24–5
Banks, Sir Joseph, 2, 4, 5, 8
barking, dog, imitated, 7, 8, 95–6
barring of feathers, 90–1
Bass, George, 5
Bass Valley, 36, 98
bathing, lyrebird, 20, 78
Batman, John, 8
beetles, as food, 15, 130, 136
Belcher, H., 114
Bill, M., 19, 26, 98
Binna Burra guest-house, 98, 127, 131
bird of paradise, 14, 15, 27
*Birds of Australia* (G. J. Broinowski), 5
*Birds of Australia* (J. Gould), 7, 80
*Birds of Australia* (G. Mathews), 49, 127
*Birds of Paradise and Bower Birds* (T. Iredale), 14,
*Bird World, The* (W. H. Davenport Adams), 5
'Bk/Bk/W', lyrebird named, 85–6
Blackall Range (Qld), 127
blackbird, English, 93, 102, 121
Black Mountain (Vic.), 106
Blood's bird of paradise, 14
Blue Mountains, 15, 104

'Blue/White', lyrebird named, 24, 78
bowerbirds, satin, 14–15, 27, 57, 80
    imitated, 99, 132, 137
bracken, areas of, 123, 124
breeding cycle, 119
    external/internal factors, 57–9
    in captivity, 104
    female, 41, 59–60
    male, 60–2
    of Albert's lyrebird, 130–1
    season, 29, 43, 119, 133
    song and, 97
    temperature and, 70–2; *see also* courtship display; non-breeding season
'Broadcast Area' (Sherbrooke Forest), 11, 24, 29, 30, 34, 102, 123
'Broadcaster, The', 4
brooding, chick, 60, 64, 71
brood patch, 59
Brush, A.H., 92
burning-off, advantages of, 124
bushfires
    and food, 123–4
    survival behaviour of lyrebirds during, 19–20, 106, 124
butcherbird, grey, 98, 102
B/W, lyrebird named, 87–8

calls,
    Albert's lyrebird's, 133
    courtship, 96, 98, 100
    Edward's lyrebird's, 136–8
    imitated, 98–102, 133
    mature lyrebird's, 97–8
    Tasmanian birds', 115; *see also* mimicry; vocalizations
camouflage, 37, 45, 136
Campbell, A.J., 6, 9–10, 22, 26, 45, 47, 53–5, 72, 80
captivity, lyrebirds in, 8, 102–10
catbird, 133
cats, feral, as predators, 25, 47
    native, 37
central nervous system (CNS), birds', 58–9, 93
'changeling', 36
chains, sound of imitated, 98
'chasey', males playing, 7, 21–3, 33
chicks, 38
    auditory signals from, 60, 95
    death of, 71
    eye of, 64
    first described, 8
    growth of, 72–3
    leaving nest, 51, 68–70, 73
    males and, 37
    newborn, 63
    post-nest development, 73–5; *see*

*also* nestlings; young lyrebirds
'Chook', lyrebird called, 24
circadian rhythm, birds', 58
'clack . . . clack . . .' call, 98, 132
Clarence River district (Qld), 130
classification of lyrebird, 14
claws, *illus.* 28, 36, 73
'clickety clickety' call, 102, 132, 137
'climbing', by chick, 73, 95
'clonk clonk' call, 98, 100, 102, 115, 117, 121, 132, 137
clucking, while mating, 131
CNS: *see* central nervous system
CNS–hypothalamus–pituitary system, 59, 62
cockatoo
    black, imitated, 99, 100, 102, 116, 120, 121, 137
    gang-gang, imitated, 100, 102
coition, 59, 60, 98, 131
'come and play' call, 102
communicating, how, 22; *see also* song; calls; vocalizations
conservationists, 112
copulation: *see* coition; mating
courtship activities, 17, 20, 24,
    Albert's lyrebird, 131–2
    calls during, 27, 29, 32, 33, 96, 98, 100
    male, 27, 29–33; *see also* displays; song; mating; preening
Coyle, Mr & Mrs J., 19, 53, 104
cradle nests, 47
'Crescent, The', lyrebird called, 11
'Crossed Lyrates', lyrebird called, 11, 24, 30
cuckoo, golden-bronze, 121
cuckoo, shrike, black-faced, 121
currawong, imitated
    black, 116, 120, 121
    pied, imitated, 96, 99, 100, 102, 120, 132, 137
currawong song dialects, 136

daily rhythm (circadian), bird, 58
'dance, the', 32, 132
Dandenong Ranges (Vic.), 9, 43, 133
Darwin, Charles, 14
death, causes of, 19, 68, 71
defecation, chick, 68
descriptions, lyrebird, first, 2–5
desertion, mother–chick, 75
dialects, bird song, 136
diet: *see* food
digit length, toe, 73
Dillon, Clem, 117
dingo, mimicking, 7, 8
discovery, lyrebird, first, 1–5
*Menura edwardi*, 135

displays
  Albert's lyrebird, 131–2
  courtship, 17, 29, 32–3, 131–2
  female threat, 42
  full-face, 32, 60
  invitation, 29, 33, 60
  mock-courtship, 29–31
  mutual, 75
  play, 22
  pre-coition, 60
  pre-courtship, 60
  sexual, 58
  wing, 24
  young lyrebird, 75–8, 96
dogs, domestic
  imitated, 24, 95–6, 98
down, chick, 60, 64–6
Double Black, lyrebird called, 78
Double-White (W/W), lyrebird
  called, 19, 77–8
drinking spots, 32, 51
'Droopy', lyrebird called, 24, 37, 41,
  45, 53

eagle, whistling, 98
Eastern Highlands, 15
Ebor district (NSW), 100
echidna (spiny ant-eater), 135
Edward's lyrebird: *see* Prince
  Edward's lyrebird
'eeeeeek' call, 97
'eeaweeaweeaw' call, 98
eggs, lyrebird
  Albert's, 131
  Edward's, 135, 136
  fertilization, 59
  incubated, 37, 55–6, 59, 70–2
  in ovaries, 19
  laying, 42, 53–4, 70
  superb lyrebird, 54–5
Elliott, A.E., 112
embryology, lyrebird, 71–2
*Emu* 11, 112, 113
endocrinal system, 38, 42, 57, 92–3,
  97, 101
environmental factors, for breeding,
  57–8; *see also* habitat
enzymes, 92–3
eumelanin, 93
evolution, how viewed, 51
excreta, chick, 68–70
exploitation, lyrebird, 8–10, *see also*
  hunting; trading
extinction, threat of, 112, 119
eyes, lyrebird's, 22, 64

Falls Picnic Area (Sherbrooke), 19,
  41, 95, 123
fantail, grey, 99, 101, 121

fauna, forest, threatened, 123
feather
  development, chick, 63–6
feathers, tail
  Albert's lyrebird, 127–30
  lyrates, 80, 83–5, 87–8
  medians, 80–5, 88–9, 90, 91
  rectrices, 80–5; *see also*
    filamentaries; tail feathers
feeding, 33
  by Edward's lyrebirds, 136
  birds in captivity, 104
  chicks, 38, 66–8, 73
  chicks learning, 73–4
female superb lyrebird
  and chicks out of nest, 73–5
  breeding cycle of, 59–60
  brooding by, 65
  courtship and, 32–3, 42, 100
  description, 4, 8, 11, 39–40
  egg-laying, 17, 22, 136
  general observations on, 13, 17
  mating, 24, 42
  mimicry by, 97
  moulting of tail feathers, 83
  nest-building by, 43–53, 136
  nesting, 37, 41, 136
  tail feathers of, 39–40, 88
  territory of, 24–5, 41
  vocal behaviour of, 96–7, 99
fertilization, egg, 42–3, 55–6
FHS: *see* follicle-stimulating hormone
fighting, lyrebirds
  inter-female, 41, 136
  inter-male, 33–6
  inter-young, 75–6
filamentary feathers, 30, 32
  *M. alberti*'s, 129
  dimensions, 83
  first, 80, 85–7, 89–91
  moulting of, 85–6
  structure, 81–2
Fisheries & Wildlife Department, 11
*Five Years in Australia Felix* (G. H.
  Haydon), 7
fledglings, 47, 73–5, 78, 95
Florentine Valley (Tas.), 113, 117,
  123–4
flute, imitated, 100
flying, lyrebird, 34, 73
follicle, feather, 89
follicle-stimulating hormone (FSH),
  58, 59, 60
food, of lyrebird, 15–17, 20, 52, 66,
  124, 130, 136
  chick's, 73
  Tasmanian, 124
forest
  chicks in, 73–5

floor, destruction, 123, 127
  litter, 124
  lyrebird, preferred, 15, 127
  management, 124
  open ground, 124
  Tasmanian, 116
foster-parenting, 38
foxes, red
  as predators, 25, 37, 47, 49, 51,
    111, 123
  lyrebirds imitating, 24, 98
  poisoning of, 25
Freestone Creek (Vic.), 124

Gemmell, A., 135
genetic factors, 58, 92, 101
George, Graham, 24
*Ghania* sp., 116
Giacomelli, H., 5
Gibb, N., 106
Gippsland, 7, 9, 15, 28, 98, 103,
  124, 133, 136
Girraween National Park, 133, 135,
  136, 137
Glenadale National Park, 53
Glenadale district (Gippsland), 15
gonadotrophins, 58
Gondwanaland, 14
goshawk, 102
graminivorous, lyrebird as, 20
'Granite Belt' (SE Qld), 6, 133–4,
  136
gray gallito (*R. lanceolata*), 14,
Groom, Arthur, 98

habitat, lyrebird
  Albert's, 127, 131
  forest preferred, 15, 118, 133
  Granite Belt, 133–4
  range of, 15, 127
Hardcastle, Mrs M., 137
Harman, Colin, 131
Harrison, Arthur, 2, 5, 62
Hastings Cave area (Tas.), 114, 118,
  120–4 *passim*
hatching, lyrebird chick, 56, 63,
  104, 106
hawks, 37, 80
Healesville, 25, 124
Healesville Sanctuary, 24, 38, 72,
  104, 106
Healesville–Warburton districts, 15,
  105, 133
hen, domestic, 62, 70–2
hearing, lyrebird's, 22–3
hissing call, 121
historical introduction, lyrebird,
  1–12
homosexual behaviour, 31

honey-eater, 99
  black-headed, 116, 120, 121
  brown-headed, 99, 102, 116, 120
  crescent, 16, 120, 121, 138
  Lewin, 133
  white-eared, 133, 137
  yellow-faced, 138
  yellow-throated, 116, 120, 121
hormones
  breeding cycle, 57–62, 92–3
  tail development, 92–3
Howe, Harry, 112, 124
Hunter, Governor John, 1–2, 5, 8
hunting lyrebirds, 8–10, 37
hybrids, lyrebird–hen, 50
hyperthyroidism, 93
hypothalamus, 58–9, 92–3

ICSH: *see* interstitial cell-stimulating
    hormone
Illawarra district (NSW), 7, 75, 103
*Illustrated London News,* 130
immature lyrebirds: *see* chicks;
    nestlings; young lyrebirds *also*
    *under* male lyrebirds; female
imitated calls *see* mimicry
incubation, egg, 55–6, 59, 63, 70–2,
    130, 136
interstitial cell-stimulating hormone
    (ICSH), 58, 60
invitation display, 29, 33, 60

'Jack', lyrebird called, 19, 24, 80
'James', lyrebird called, 24
'Joe', lyrebird called, 104
Juhn, M., 92, 93

Kaup, Professor, 8
King, Governor, 4
Kinglake district (Vic.), 9
Kinglake National Park, 9
koala, 106
  grunting imitated, 98
kookaburras
  as predators, 25
  mimicked, 32, 96, 99, 100, 101,
    102, 114, 120, 121, 133, 137
  stealing food, 52
Kurmond (NSW), 26

Lamington National Park, 98, 127,
    131
'Landslide, Female, The', lyrebird
    called, 24, 41
Latham, John, 6, 13
Leadbeater, J., 8
learning, lyrebirds', 37, 51, 73, 95–6
legislation, protecting lyrebirds, 9
legs, lyrebird, 28; *see also* claws

Linnean Society of London, 4, 125
logging, Tasmanian, 117–18, 124
log-runner, spine-tailed, 132
longevity, lyrebird, 17–18, 105
luteinizing hormone, 59, 60
Lyndon, Mrs Ellen, 111
lyrate feathers, 80, 82 *chart,* 85
  Albert's lyrebird's, 127
  Edward's lyrebird's, 135
  female's, 39, 83
  generational development, 87–9
  immature male's, 84
  mature male's, 84
  pigmentation of, 93
lyrebird: *see Menura*
'lyrebird', as vernacular name, 6

McKenzie, Sir Colin, 106
MacPherson Range, 127, 133
magpie, white-backed, 101–2
  black-backed, 103
Main Dividing Range, 133
male lyrebird, superb
  breeding cycle, 60–2
  colouring, 26
  courting and mating, 17, 24, 31–2
  feathers of, 83–4
  fighting, 33–6
  general observations, 13
  immature, 22, 30, 31, 83–4, 76–8
  relationship with chick, 75
  nesting behaviour, 36–8
  tail development, 88–91
  tail moult, 17, 83–4; *see also*
    displaying; mounds
mannikin (bird), 27
Mason, Kevin, 26, 38
mating, 17, 24, 31, 32, 43 *illus.,* 58,
    60–2, 80, 98, 131; *see also*
    coition, courtship displays
maturity, lyrebird, when reached, 39,
    80, 91
Maydena district (Tas.), 121–3
mechanical sounds, imitated, 98,
    100, 132
melanin, black, 93
Melbourne Exhibition (1854), 8
memories, lyrebird's, 51
*Menura*
  *Alberti,* Bonaparte (Prince Albert's
      lyrebird), 125–37
  *Edwardi* (Prince Edward's
      lyrebird), 6, 53–4, 70, 129,
      133–8
  *novae-hollandiae,* 6, 54, 135–6
    *intermedia,* Mathews, 6
    *novae-hollandiae* Latham, 6
    *victoria* Gould, 6
  *superba,* 4, 6, 125–7; *see also* male

  superb; female superb
*Menurae,* ancestors of, 14–15
*Menuridae,* 14
mimicry
  Albert's lyrebird, 132
  analysing, 100–2
  as talk?, 24
  by female, 97
  by male, 98–101
  chicks learning, 37, 95–6
  ease of, 118
  Edward's lyrebird, 137–8
  first accounts, 5, 7–9, 11
  purpose, 99–100
  young birds developing, 75–6; *see*
    *also* individual sounds
Misery Plateau (Tas.), 117, 122
Mistake Range (Queensland), 127
Mitchell Gallery, Sydney, 5
Monbulk State Forest (Vic.), 10
monogamy, 24
moulting
  head and neck, 17, 78
  shock moult, 83
  tail moult, 17, 78, 82–3, 85–93
  tail moult of female, 78, 83
mounds, display, 6, 27–9
  Albert's not using, 131
  dimensions of, 28
  Edward's, 136
  first accounts, 7, 11
  ground preferred for, 123
  male displaying on, 17, 22, 32–3,
    60
  number of, 29
  refurbishing, 29
  young lyrebirds on, 75–6, 96
Mount Buffalo district (Vic.), 15,
    28–9, 99
Mount Field National Park (Tas.), 29,
    112–15, 119, 121
Mount Field lyrebirds, 114, 117,
    118–21
Mount Toole-be-wong (Vic.), 26, 47
multiple-laying, egg, 54–5
music, lyrebird creating own, 102;
    *see also* song
musk (*Oleria argophylla*), 15, 117
mutual interaction, young birds,
    76–8

names, lyrebird, finding, 2, 4–8; *see*
    *also* individual names
National Museum (Melbourne), 54,
    72, 79, 131, 135
'native pheasant', 2
naturalists, 10
Neerim district (Gippsland), 9
nesting behaviour

female: *see under* female superb
  lyrebird
  in captivity, 106
  of Albert's lyrebird, 130
  male, 36–8
nestling, lyrebird
  calls of, 95
  eye of, 64
  feather development of, 64–5
  feeding, 66–8
  leaving nest, 68–70
  newborn, 63
  temperature of, 64, 71
  throat coloration, 78 *see also*
    chicks; young lyrebirds
nests, 25
  Albert's lyrebird, 130
  design and construction of, 27, 36,
    41, 43–53
  destruction of, 41
  Edward's lyrebird, 136
  first collection of, 8
  ground preferred for, 123
  platform, 47, 53
  sanitation of, 68
  temperature in, 64
  time of building, 53–4, 59
*Nests and Eggs of Australian Birds*
  (A.J. Campbell), 9
neural system, 57; *see also* CNS
New England National Park (NSW),
  100, 135
New Holland Menura (*Menura
  novae-hollandiae*), 6
Newton, Yolande, 123
noises, vocal, lyrebird: *see*
  vocalizations
non-breeding season, 29, 31, 33, 41,
  97, 136
Nothofagus: *see* Antarctic beech

Odell's Gully, 25, 37, 41, 49, 53
oil gland, tail base, 20
'Old Changeling, The', lyrebird
  called, 87
*Olearia argophylla* (musk), 15, 51,
  73, 95, 117, 123
'One-Eye', lyrebird called, 36
O'Reilly's Guest House, 127, 131
origins, early, lyrebird, 14–15
ornithologists: *see* individual names
osteology, 14
Otway Ranges (Vic.), 15
outposts, original lyrebird's, 6; *see
  also* range
ovary activation, 58, 59
ovulation, 59, 60, 62
owls, 37, 47
  boobook, imitated 98

Palmer, J.S., 87, 103
pandanni plants, 116
papers, scientific, 4–5
Paradise riflebird, 133
parrot, King, 133
passeres (perching birds), 14
peacock, 80
  lyrebird's tail like, 2
perches, lyrebird
  favourite, 20, 51
  for chicks, 73, 95
perching birds: *see* passeres
phaemelanin, 93
pheasant, 14, 56, 80
'pheasants', lyrebirds called, 2, 6, 8,
  13
Phillip, Governor, 4
photographers, lyrebird, 10–11, 132
pigmentation, feather, 78, 93, 127
pilot bird, 15, 33, 99, 100, 102
pineal gland, 58
pituitary, 58, 59, 92
plain feathers, 88, 89, 90–1
plaintail, 29, 30, 31, 33, 36, 77
platforms
  display, 131–2
  nesting, 47, 53
platypus, 106, 135
play, lyrebird, 21–2
Pleistocene glaciation, 15
Plenty Ranges (Kinglake, Vic.), 9
plumage, adult, 80; *see also* feathers
polygamy, 24–5, 119, 120
Pomaderris, 117, 123
population, lyrebird, 15, 123–4, 127
  decline in, 49, 123
  density, 27, 127
  growth in, 119–20
Port Phillip District (Vic.), 6, 7
predators, lyrebird, 19, 23, 25, 37,
    49, 83, 95, 120
  and nests, 49, 51, 68
preening, 20, 22
Price, John, 2
Prince Edward's lyrebird (*Menura
    edwardi*), 6, 53, 100, 129
  behaviour, 136–7
  designation, 135
  discovery, 135
  introduction, 133–5
  range of, 135, 136
  song, 137–8
  tail feathers, 135
'Prince of Mocking Birds', lyrebird as,
  94
progesterone, 59
prolactin, 59–60
protein, for feathers, 71, 92
*Pteroptochos* (South America), 14

Queensland, 15
  lyrebirds in, 125–37
Queen Victoria's lyrebird, 6

range of habitat, lyrebird's
  Albert's, 127
  Edward's, 134–6
  superb, 15, 127, 135
  Tasmanian, 121–4
ravens, 25, 71, 101
rectrices, feather, 80–1, 83–5
'Red', lyrebird called, 24, 87–9
  *passim,* 96
'Red/Blue', lyrebird called, 19, 76,
  87, 88
'R/Y', lyrebird called, 87, 88
reproduction cycle: *see* breeding
  cycle
*Rhinocryptidae* (family), 14–15
  *lanceolata,* 14
  *Pteroptochos,* 14
Richmond River, 125, 127, 130, 131
robins, 33, 45 *illus.*, 57,
  yellow, imitated, 99, 100, 101,
    102, 133, 138
*Romance of the Lyrebird* (A.H.
    Chisholm), 6
roost, overnight, 73, 78
rosella
  crimson, 99–102 *passim,* 120,
    133, 137
  green, 116, 120, 121
Rowe, G., 113, 114, 121
Royal Australian Ornithologists
  Union, 112
Royal Automobile Club of Victoria,
  106
Royal Melbourne Zoo, 104, 109
ruffs, 27
Russell Falls (Tas.), 113, 121

sanitation, nest, 68
sassafras (*Atherosperma
    moschatum*), 15, 117
Scenery Preservation Board (Tas.),
  113
scientific papers, first, 4–5
'scissors . . . scissors' call, 98, 100,
  102, 121, 137
scratching, litter, 136
scrub-birds, 14, 15
  noisy (*A. clamosus*), 14
  rufous (*A. rufescens*), 14
scrub wren, 15, 33, 121
seasons, lyrebirds and, 17, 29, 32,
  59, 70
self-defence, 95; *see also* alarm call
sex
  hormones, 58, 60, 92–3, 97

organs, 57, 60, 92–3

sexual activity, periods of, 30–3; *see also* mating; courtship

sexual maturity, 39, 80, 91, 93

Sherbrooke Forest, 11, 19, 20, 24–5, 28, 47, 49, 51, 53, 62, 70, 98, 99, 101–2, 114, 123, 133, 135

Sherbrooke Gully, 25, 41, 53, 66

Sherbrooke Survey Group, 11, 24, 25, 54, 79, 106

*Shooting in Both Hemispheres* (J.D. D'Ewes), 8

shrike thrush, grey, 116

shrike tit, eastern, 99, 102

'Silvertail', lyrebird called, 29

*Silvertail – The Story of the Lyrebird* (Watson), 29

silver-eye, imitated, 99, 101, 102

singing, lyrebird: *see* song

'Singing Hen', lyrebird called, 97

'Single Red' (R), lyrebird called, 19, 101

Sir Colin McKenzie Sanctuary (Healesville), 38, 104, 105

*Sketch of Natural History in Australia, A* (F.G. Aflalo), 9

skins, lyrebird, 2, 4, 79

sleeping, lyrebird, 108–9

'snaking of the head', 22

social behaviour, young lyrebird, 77–8; *see also* varieties of

song, lyrebird, 11, 33
 Albert's lyrebird's, 130, 132–3
 bathing, 20
 compared, Tas. & Vic., 114
 courtship, 17, 27, 29, 32, 33, 96, 98, 100
 duration of, 117
 Edward's lyrebird's, 137–8
 group, 96
 loudness of, 100, 130
 pre-courtship, 60
 Tasmanian lyrebird's, 120–1
 specific calls, 98
 survival and, 101
 young lyrebird's, 95–6, 101–2, 121; *see also* mimicry; vocalizations; calls

Southern Highlands (NSW), 2

spaghnum moss beds, 116

species, lyrebird, 6, 14, 135
 Edward's as?, 135
 superb & Albert compared, 125–30

specimen, lyrebird, first, 2, 4–6, 8–9, 103, 127

sperm production, 60

'Spotty', lyrebird called, 11, 17, 19, 22, 24, 29, 33, 37, 51, 123

'Spotty Junior' (YY), 30, 76–7, 88, 96

Springbrook Plateau (Qld), 127

squawking, call, 22, 96, 97

stalking, lyrebirds, 7, 10, 22

Stanthorpe district (Qld), 135

steam train, imitating, 98

Stephenson, Dr, 127

Strange, Frederick, 125, 127

strychnine, 25

sub-species, 6

Sundown National Park, 135, 136

superb lyrebird
 breeding cycle of, 57–62
 compared Albert's, 126–30
 compared Edward's, 135
 early history, 1–12
 female, 56
 general observations on, 13–27
 in captivity, 103–10
 in Tasmania, 111–24
 male superb, 26–38
 song of, 94–102
 tail of, 79–93
 young, 63–78

survival, factors in, 51, 58, 64, 95, 101

Sydney Museum, 125

tail feathers
 and courtship, 32–3
 and nesting, 37
 chick's, 66
 complete, 91, 130
 displaying, 29, 32
 during coition, 62
 endocrinal system, role in, 92–3
 female, 39–40
 general structure, 79–82
 generational development, 85–91
 male, 79–93
 moulting and regrowth, 17, 29, 31, 33, 62, 75, 80, 82–91
 selling for profit, 9
 while resting, 37

*Talitris* sp., 15, 66

'talking' lyrebird, 24
 chick to mother, 95

taming, lyrebirds, 103, 105

Tamborine Mountain (Qld), 127

tapaculos *(Rhinocryptidae)*, 14

tape-recording, 101, 114, 116, 118

Taronga Park Zoo (Sydney), 104

tarsus length, 73

Tasmania, 14, 28, 100
 lyrebirds in, 111–24

Tasmanian lyrebird
 birds imitated by, 121
 distribution, 121–4
 habitat, 115, 117, 124
 introduction of, 100

population, 119–20
 range of, 121–4
 song of, 120

Tasmanian National Parks & Wildlife Service, 115

temperature, effect of,
 on breeding, 70–2
 on feather development, 64

territorial calls, 96, 98, 99, 100, 102, 121, 133, 136, 137

territory
 Albert's lyrebird, 130
 female, 24, 41
 male, 24, 27

testes, activation of, 58, 60

testosterone, 60

thornbill, 133
 brown, 137

threat display, female, 42

throat coloration, 78

thrush, 14
 grey, 66, 96, 100, 101, 116, 121, 132, 133, 137
 song of, 136
 Tasmanian, 136

thyroid gland, 60, 92

thyroxine, 92–3

Tibinbilla (ACT), 15, 28, 66, 70, 137

'Timothy', lyrebird called, 11, 17, 22, 33–4, 123

Toolangi–Warburton district (Vic.), 112, 114, 124, 130

trees rubbing, imitated, 137

Tubb, J.A., 114

Turanga Creek, 127

Tweed River, 127

Tyenna River (Tas.), 115

Tyenna Valley (Tas.), 121

tyrosine, 93

'ugh . . . ugh' call, 98

Uralba, 127

Victoria, early lyrebird accounts, 7–8

Victorian government, 106

vision, lyrebird, 22–3

vocalizations, 29
 chicks, 60, 94–5
 courtship, 32–3
 exciting female, 100–2
 fighting, 35–6
 listening, 22
 mature male, 97–8
 mother's, 95–7
 while playing, 22, 102; *see also* song; calls; mimicry

*Vulpes vulpes* (red fox), 25

Walkerville district, 111

Waller, Fred, 53
'Wanderer, The', lyrebird called, 11, 34
*Wanderings of a Naturalist in
    Australia, The* (G. Bennett), 103
Wandong district (Vic.), 15
Warburton district (Vic.), 20, 124
'warrrrh' call, 98, 102
wattlebird, imitated, 100, 102, 138
    red, 120
    yellow, 120, 121
Wattle Walk, Sherbrooke Forest, 32,
    41, 123
Weld Valley (Tas.), 121
Westernport district (Vic.), 7
Westron, T., 115
whipbird, 96, 98
    imitated, 99, 100–1, 114, 119,
        120, 121, 133, 137–8
White, H.L., Collection, 54, 55, 131,
    135
white settlement, impact of, 112
'whisk whisk' call, 97
whistle, human, imitated, 98, 118
whistler, imitated
    golden, 99, 100, 102, 120–1, 138
    olive, 99, 115, 120, 121
'White/Red', lyrebird called, 39
Wilcox, J.F., 130
Wilson, John, 1–2
Wilsons Promontory, 15, 111
wings, lyrebird, 24, 32, 66
wombats, 135
    burrows, 20, 36, 37
wonga pigeon, imitated, 103
wrens, 80
    scrub, 102, 137
'WW', lyrebird called, 77–8, 87–8,
    92 *illus.*
'WWR', lyrebird called, 86

yearling chick, 75
young lyrebirds, 63–78
    Albert's, 130
    display and song, 75–8
    Edward's, 136
    eye of, 64
    feather development of, 64–6
    feeding, 66–8
    forest life of, 73–5
    general observations, 17
    growth of, 72–3
    identifying sexes of, 97
    in nest, 63–4
    leaving the nest, 68–70
    mutual interaction of, 76–8, 136
    nest sanitation and, 68
    social behaviour of, 77–8
    tail feathers of, 83
    temperature, effect on, 70–2

wings of, *see also* chicks; nestlings
'Y/Y' (Spotty Junior), 30, 76–7,
    88, 96

zig-zag flight, 51–2, 75
'Zoe', lyrebird called, 104
Zoological Society of London, 86–7,
    103, 104
zoos, lyrebirds in, 103–6

# Author index

Adams, W.H., 5
Aflalo, F.G., 9
Ahlquist, J.E., 15

Bartlett, A.D., 14, 104–5
Becker, L., 8, 103, 104
Bellairs, R., 71
Bennett, G., 87, 103–4, 125
Bill, M., 19, 26, 98
Bradford, S.L., 53
Broinowski, G.J., 5
Brunner, H.L., 25
Brush, A.H., 92

Cabanis (quoted by Feduccia and
    Olson, q.v.), 8, 14
Campbell, A.G., 49, 72, 80
Campbell, A.J., 6, 9–10, 22, 26, 45,
    47, 53–5, 72, 80
Chandler, R.C., 36, 54, 55
Chisholm, A.H., 6, 55, 104–5, 127,
    130, 135
Collins, D., 4
Coman, B.J., 25
Cook, L.C., 119
Curtis, H.S., 127, 131, 132

Davies, T., 4, 6, 14
Denton, Sherman F., 9
D'Ewes, J., 8–9
Dorst, J., 58

Edwards, A.V., 49
Eisner, E., 60
Eyton (quoted by Feduccia and
    Olson, q.v.), 14

Feduccia, A., 14
Fleay, D., 37–8, 72, 80, 86, 105,
    112
Frieden, E.H., 60
Frith, H.F., 66, 70

Gillan, Miss, 80
Glegg, W.E., 90
Godfrey, F.P., 19, 24, 80
Gould, J., 6, 7, 14, 21, 22, 78, 80,
    125, 127, 130, 131
Gray, A., 49, 80
Grey, J.E., 86–7, 103

Halafoff, K.C., 101
Hartshorne, C.E., 102
Haydon, G.H., 7–8

Heine (quoted by Feduccia and
    Olson, q.v.), 14
Hinde, H.A., 60
Hindwood, K.A., 53, 55
Höhn, E., 60, 92
Howe, F., 20, 47

Jackson, S.W., 53, 127, 131
Juhn, M., 92, 93

Keast, A.J., 130–1
Kenyon, R.F., 11, 14
Kitson, A.E., 20, 49, 112, 136

Lawler, W.G., 132
LeSouef, D., 36, 55, 104
Lewin, R., 92
Leycester, A.A., 127, 130–2
Lill, A., 25, 37, 54, 56, 64, 70, 71–2
Lillie, F.R., 92, 93
Littlejohns, R.T., 11
Lloyd, J., 25
Loyn, R., 124

Mack, G., 50
Mahan, J.C., 53, 80, 104
Maroney, D., 78
Marshall, A.J., 58, 96
Mathews, G.M., 6, 49, 78, 127, 131
Milligan, A.W., 49–50
Moreau, R.E., 14

Newton, A., 14
Nicholls, B., 98

O'Donoghue, J.G., 49
Olson, S.L., 14

Portman, A., 68
Pratt, A., 24, 80

Rawnsley, H.C., 45, 75, 98
Reilly, P.N., 25
Ridpath, M.G., 14
Roberts, S., 6, 53, 55, 127, 135, 136
Robinson, F.N., 15, 66, 70, 133
Rowley I., 27, 71

Serventy, D.L., 14
Sharland, M.S., 11, 114, 115
Sibley, C.G., 14, 15
Smith, L.H., 79
Stone, A.C., 55

Terres, J.K., 31
Thomson, D.F., 68
Thorpe, W.H., 95
Tregellas, T., 10–11, 25, 41, 47, 70,
    80, 98, 113, 123

Vidler, E., 137
Voitkevich, A.A., 60, 92, 93

Wall, L.E., 114–15
Wallace, J., 54
Ward, J.E., 25, 56, 70
Watson, G.E., 93
Watson, I.M., 12, 29, 62
Welty, J.C., 58
Westerkov, K., 56
Wheeler, W.R., 114
Williams, F.J., 8, 80
Witschi, E., 58
Wood, H.B., 90–1
Wyld, J.A., 92